HAS FEMINISM CHANGED SCIENCE?

HAS FEMINISM CHANGED SCIENCE ?

Londa Schiebinger

HARVARD UNIVERSITY PRESS

Cambridge, Massachusetts
London, England
1999

Library of Congress Cataloging-in-Publication Data

Schiebinger, Londa L.
  Has feminism changed science? / Londa Schiebinger.
    p.   cm.
  Includes bibliographical references and index.
  ISBN 0-674-38113-0 (alk. paper)
  1. Women in science. 2. Women scientists. 3. Feminism.
  I. Title.
  Q130.S29 1999
  508.2--dc21
  98-42423

For my parents, with love

Contents

Acknowledgments

THIS BOOK was much harder to write than I ever imagined. I am especially grateful to the scholars who took time to guide me through the intricacies of their fields. Linda Fedigan provided crucial correctives concerning primatology along with insights into women's status in the discipline. Scott Gilbert's humor and good cheer led me through some dark and otherwise mysterious areas of biology. Amy Bug kept me from losing nerve in my conclusions to Chapter 9; and Catherine Kallin, Katherine Freese, and Elizabeth Simmons invited me to the Aspen Center for Physics, where I enjoyed not only the rarified air of the Colorado Rockies but the pleasures of frank discussion concerning gender in physics. Catherine and her husband John Berlinsky kindly read Chapter 9 and helped with several important points. Contact with Adrienne Zihlman reminded me, the historian whose subjects are usually dead, that talking to historical figures can be extremely profitable. Adrienne also supplied useful documents and viewpoints. Nancy Krieger redirected my interpretation of aspects of medical history and offered other valuable insights. Mary Golladay at the National Science Foundation graciously answered questions about statistics and sent along tomes of information.

Many of my colleagues from science departments at Pennsylvania State University amiably suffered through interviews probing their lives, their career advances, and "changes women had made in science," as I was then improperly phrasing it. These interviews became a radio series on women and minorities in science. I also had many opportunities to discuss these issues with working scientists at the Space Telescope Science Institute in Baltimore, Cornell University, the University of Chicago, Oregon

State University, Harvard University, Georg-August-Universität in Göttingen, Karlsruhe Universität, the Royal Danish Academy of Letters and Sciences in Copenhagen, the University of Lund (Sweden), and elsewhere.

A special thanks to Natalie Davis, who made possible a year at Princeton University, where this project began. Thanks, too, to Thomas Laqueur, Lorraine Daston, Roy Porter, and Everett Mendelsohn, who have been generous supporters of my projects over the years. I am indebted to Ilse Costas and Regine Kollek, who kept me up-to-date on developments regarding women and science in Germany and facilitated my lengthy visit to Germany in 1995. Thanks also to Elke Kleinau, who arranged a room for me in her institute at the University of Hamburg, where much of this book was written. Research support was provided by the National Science Foundation, the Deutsche Forschungsgemeinschaft, and the Office for Research and Graduate Studies at Penn State.

I am grateful to friends and colleagues who nodded appreciatively each and every time I told them the manuscript was finished: Nancy Brown, Mary Pickering, Claudia Swan, Susan Squier, Gillian Hadfield, France Cordova, Amy Greenberg, Sandra Harding, Margaret Jacob, Joan Landes, Dorothy Nelkin, Bonnie Smith, and Lindy Brigham. I also owe a great deal to my fine research assistants, Anne Demo and Linda Lasalle. Thanks, too, to my editor, Elizabeth Knoll, for her helpful comments.

Finally, Robert Proctor remains a loving source of inspiration and support.

HAS FEMINISM CHANGED SCIENCE?

Introduction

FEMINISM has brought some remarkable changes to science. Who, just a decade ago, could have predicted that the chief scientist at NASA would be a woman, or that the Secretary of the Air Force would be a female professor of engineering? Who would have expected to see *Science,* the nation's premier scientific journal, debating whether a "female style" exists in science, or Marie Curie, once shunned by the prestigious Parisian Académie des Sciences, exhumed and reinterred in the Panthéon, the resting place of heroes like Voltaire, Rousseau, and Victor Hugo?[1]

The question of gender in science is pursued by scholars from many disciplines and with widely varying perspectives. Historians study the lives of women scientists in the context of institutions that for centuries held women at arm's length; sociologists focus on women's access to the means of scientific production; biologists scrutinize how scientists have studied women; cultural critics explore normative understandings of femininity and masculinity; philosophers and historians of science analyze the influences of gender on the content and methods of the sciences.

In this book I summarize and analyze these sundry scholarly approaches. What one gains in breadth, of course, one loses in specificity. Even for specialists, however, it is useful to take stock, to reconsider the relationships among various lines of investigation and argumentation, to identify what has been done and done well, and to highlight questions that remain to be answered. Within gender studies of science, I am also trying to effect a shift away from abstract critique toward the more positive task of asking what useful changes feminism has brought to science. For long enough we have asked what science is doing wrong. It is time to look at what gender studies can offer in the form of new perspectives, new research projects and priorities.

The literature on gender and science is scattered across the academy and often written in the dialect of a particular discipline. Philosophers and historians of science have made gender and science an area of expertise; academic positions are now being established in this field of study. But knowledge that has been so assiduously culled from archives is still little known among scientists, even sometimes among those with a keen interest in the topic. Lack of time and the rigors of the laboratory are clearly reasons. But, more than that, this literature is sometimes difficult—written in the high and sometimes esoteric language often required of academic humanists for advancement within their profession. At one level, I see this book as a translation project, an attempt to make clear to readers of diverse backgrounds and interests the important issues concerning the place of women and gender in science.

The current "science wars," as the often-childish tussles between scientists and their critics are unfortunately called, provide a certain measure of the successes of feminism in science. I was shocked to read in Paul Gross and Norman Levitt's *Higher Superstition* that "the only widespread, *obvious* discrimination today is against white males," but I was more surprised at the extent of our agreement. Feminists and some of their most vocal opponents agree that women should have a fair chance at careers, inside and outside academic life. We agree that some women thinkers have been rightfully restored to their place in history. We agree that the "record of science, until recently, is—in its social aspect—tarnished by gender-based exclusions." We agree further that "baseless paradigms" in medicine and the behavioral sciences have been pretexts for subordinating women. "All this is beyond dispute and generally recognized," Gross and Levitt claim, even among political conservatives. This depth of agreement marks an extraordinary change for women, who were admitted to American and European universities only about a century ago, admitted to graduate programs even later, and told as recently as 1950 that women need not apply for professorships in biochemistry. By this measure, it seems we have all become feminists.[2]

One area of disagreement remains, however, and here Gross and Levitt speak for many in proclaiming that "there are as yet no examples" of feminists' having uncovered sexism *in the substance of science*. One reason for this disagreement is that Gross and Levitt focus on feminist historians and philosophers of science and overlook the contributions of scientists, many of whom not only apply feminist insights in their work but have contributed to feminist theory and practice. It is primatologists themselves, for example—not academic outsiders—who provocatively claim that primatology is a "feminist science." Whether or not one considers this claim justified, feminist interventions have remade foundational paradigms in the field. Nonhuman females are no longer seen as docile creatures who trade sex and reproduction for protection and food, but are studied for their own unique contributions to primate society. As we shall see, feminism has brought changes to other fields of science as well.

Blind Alleys

Feminism is a complex social phenomenon and, like any human endeavor, it has suffered its share of misadventures and traveled down a number of blind alleys, as can be seen in the trials and tribulations of liberal feminism, long the leading form of feminism in the United States and much of Western Europe.[3] Who, these days, is not in favor of equal opportunity for women, or, to put a label on it, who is not a liberal feminist?

Since Mary Wollstonecraft's vigorous call for equality in her *Vindication of the Rights of Woman* (1792), liberal feminism has informed major legislation guaranteeing women equal education, pay, and opportunity (the Equal Pay Act of 1963; Title IX of the Education Act Amendments of 1972; the Equal Opportunity Employment Act of 1972). It is also the guiding principle of the much maligned doctrine of affirmative action, which accelerated women's entry into the professions. Liberals generally see women as the in-principle equals of men—everything else being equivalent—and therefore strive to provide women with the skills and opportunities to make it in a man's world. Feminism at this level has made such an impact that most people no longer even think of these issues as "feminist."

While liberal feminism has served women well, it also has led into certain blind alleys. In the attempt to extend the rights of "man" to women, liberals have tended to ignore gender differences, or to deny them altogether. For all practical purposes, so the thinking goes, women think and act in ways indistinguishable from men. Only women have babies, but childbirth is supposed to take place exclusively on weekends and holidays, not to disrupt the rhythm of working life. Liberal feminists tend to see sameness and assimilation as the only grounds for equality, and this often requires that women be like men—culturally or even biologically, as when the U.S. Army introduces the "Freshette System," a plastic spout designed to give women equal opportunity to urinate while standing in the field.[4]

A second problem with liberal feminism (also called "scientific feminism," "feminist empiricism," or "equality feminism") is that it seeks to add women to normal science, leaving the latter unperturbed. Women are supposed to assimilate to science rather than vice versa; it is assumed that nothing in either the culture or the content of the sciences need change to accommodate them.[5]

In the early 1980s feminists began developing what is sometimes called "difference feminism," which embraced three basic tenets. First and foremost, difference feminism diverged from liberalism in emphasizing difference, not sameness, between men and women. (It differed from the older and more deeply entrenched tradition of biological determinism by claiming that women were fundamentally different from men by dint of culture, not nature.) Difference feminism also tended to revalue qualities that our society had devalued as "feminine," such as subjectivity, cooperation, feeling, and empathy. And the new strand of feminism argued that, in order for women to become equal in science, changes were needed not just in women but also in science classrooms, curricula, laboratories, theories, priorities, and research programs.

The philosophical roots of difference feminism can be traced to the nineteenth century, when advocates such as the German writer Elise Oelsner taught that the "superior nature of women" could reform science by directing knowledge away from the pursuit of power and toward greater equality, freedom, and fraternity for all humankind. Oelsner did not believe that feminine qualities belonged to women alone. For her, "the eternal feminine" had animated Jesus, Plato, and Schiller, men whose lives displayed supposedly feminine values—"a world-conquering virtue, readiness to self-sacrifice, warmheartedness and devotion." In our own century the psychotherapist Bruno Bettelheim held that a "specific female genius" could make valuable contributions to science.[6]

More recently the focus on culturally specific "feminine" characteristics has prompted claims that women have distinct "ways of knowing"—including "caring" (Nel Noddings), "holism" (Hilary Rose), and "maternal thinking" (Sara Ruddick)—which purportedly have been excluded from the practices of the dominant forms of science. Carol Gilligan maintained that women speak "in a different voice" when making moral judgments, that they value context and community over abstract principles. Mary

Belenky and her colleagues, in the influential book *Women's Ways of Knowing,* suggested that women use connected knowledge, contextual thinking, and collaborative discourse rather than "separate" knowledge that privileges impersonal and abstract rules and standards.[7]

The value of difference feminism has been to refute the claim that science is gender neutral, revealing that values generally attributed to women have been excluded from science and that gender inequalities have been built into the production and structure of knowledge. But difference feminism, especially when vulgarized, can be harmful to both women and science. As postmodernists from Donna Haraway to Judith Butler have pointed out, difference feminism too easily posits a "universal woman." Women have never constituted a tightly knit group with common interests, backgrounds, values, behaviors, and mannerisms, but instead have always come from distinct classes, races, sexual orientations, generations, and countries; women have diverse histories, needs, and aspirations.[8]

Difference feminism has also tended to romanticize those values traditionally considered feminine. The study of the historical construction of gender differences can provide an opportunity to understand what scientists have devalued and why; it should be recognized, however, that in cultures where women are subordinate, the celebrated "feminine" or "women's ways of knowing" usually represents little more than the flip side of culturally dominant practices. In romanticizing femininity, difference feminism does little to overturn conventional stereotypes of men and women. Today's much-touted "holism," for instance, is not unique to women and often has little to offer them. Katharine Hayles points out that the incorporation of the purportedly "feminine" (sometimes improperly identified as feminist) principles of nonlinearity and turbulence into chaos theory, for example, has done nothing to increase the number of women among that theory's practitioners.[9] While so-called feminine traits may sometimes serve as tools for critique, perhaps by allowing us to see aspects of nature that have been misunderstood or neglected, they cannot be expected to ground an entirely new kind of science.[10]

Another blind alley has been the search for women's distinctive "ways of knowing." Empathy, for example, has been credited with advancing primatology. In the mid-1980s Thelma Rowell of the University of California at Berkeley suggested that it was "easier for females to empathize with females," and that empathy was therefore "a covertly accepted aspect of primate studies."[11] Jane Goodall's great empathy for chimpanzees and research subjects more generally was thought to derive from the fact that she is female. Involvement of women in primatology was said to have brought about a revolution in the way we look at animal behavior: before the 1950s, so the story goes, primatologists caught only glimpses of the animals they sought to study. Later, in the 1950s and 1960s, primatologists (among them Japanese men) devised ways to live among the apes and, as a result, were able to observe that chimpanzees make tools, a discovery that redefined what it meant to be human.

Evelyn Fox Keller's writings about the cytogeneticist Barbara McClintock were taken as evidence that women had their own distinctive methods of research. Though McClintock was not a feminist, Keller's portrayal of McClintock's "feeling for the organism" captured the popular imagination, becoming an icon for a supposed "feminine" or, at times, even a "feminist" science. According to Keller's account, McClintock exuded a close affinity for her research subjects, listening to "hear what the material has to say to you . . . [to] let it come to you." Keller's notion of this feeling for the organism is more complex than generally understood, based as it is on an appreciation of individual difference rather than any essentialist projection of gender.[12]

Keller never argued that women—as a class of humans—employ distinctive research methods. But some feminists do argue this, asserting that women scientists tend to be holistic and integrative thinkers who, as a result of their socialization, are less satisfied with reductionist principles of analysis than men are. Linda and Laurence Fedigan have suggested along these lines that "the values traditionally defined as feminine may lead women to be generally more persistent and patient, willing to wait for the material to speak for itself rather than forcing answers out of it, and envisioning themselves as more connected to the subject matter than in control of it."[13]

Donna Haraway has noted that the methodological claims for empathy validate the feminine side of the classic dualism between feeling and objectivity, without reworking the relationship. Women have long been considered closer to nature than men. Louis Leakey reportedly sent Jane Goodall into the field because he believed women were especially patient and perceptive. Leakey commented to Sarah Hrdy in 1970, "You can send a man and a woman to church, but it is the woman who will be able to tell you what everyone had on."[14]

Efforts to refashion science by adding traditionally feminine traits can be appealing: they create immediately life-affirming projects, alluring visions of how things could be different. The attempt to attach the good and the beautiful to women, however, can unnecessarily alienate sympathetic men. Stephen Jay Gould, for example, objected to Keller's claims for McClintock's "feeling for the organism," maintaining that male scientists also "empathize" with their subjects and that little in the supposedly "feminist" method is distinctive to women.[15] Easy stereotypes concerning women and "feminine" qualities can prove needlessly divisive.

There are, to be sure, alternative methods of conducting research, but these are not directly related to sex or presumed womanly traits. In many instances feminism has been advanced through the use of standard methods of scholarship. A historian, for example, may ask new questions but answer them by using standard historical methods, such as archival research, textual analysis, demography, and comparison of evidence. Or a historian may devise new methods to answer new questions. When attempting to make visible the lives of women of a class that did not leave letters, diaries, or other written records, a historian may have to search out indirect sources, such as court records, that provide some glimpses into the lives of silkmakers, brewers, prostitutes, or midwives. These new methods may allow us to look at questions of gender, but the methods themselves are derived from long years of experience in archives and work with historical materials, and not from some set of allegedly feminine qualities.

So, too, in primatology. New methods encouraged the inclusion of formerly ignored research subjects—both females and low-status males. But again, these methods had no special attachment to qualities typically considered feminine (or masculine). In the 1970s Jeanne Altmann drew attention to representative sampling methods in which all individuals, not just the dominant and powerful, were observed for equal periods of time. (Primatologists had previously used "opportunistic sampling," merely recording whatever captured their attention.) Representative sampling required that primatologists evaluate the importance of events by recording their frequency and duration. Commonplace events such as eating, grooming, and lolling thus claimed their place next to the high drama of combat and sexual encounters, allowing for a more nuanced and egalitarian vision of primate society.[16]

Sampling methods are quantitative. Some feminists criticize quantitative methods on the grounds that quantification can miss or obscure the messy stuff of natural and social experience. In this instance, however, quantification was good for feminism, providing systematic sampling techniques that allowed primatologists to document the contributions of females to particular groups. Feminists need to avoid glorifying or con-demning methods yanked from context: too quick a dismissal of quan-titative methods limits a scholar's ability to collect and interpret data from a variety of useful perspectives. In many areas of science, as in the hu-manities, quantitative and qualitative studies can complement each other.

Women's historically wrought differences from men, then, cannot serve as an epistemological base for new theories and practices in the sciences. There is no "feminist" or "female" style ready to be plugged in at the laboratory bench or the clinical bedside.[17] Feminist goals in science will not be realized through the invocation of cliché-ridden principles drawn from a mythical "lost feminine." It is time to move away from conceptions of feminist science as empathetic, nondominating, environmentalist, or "people-friendly." It is time to turn, instead, to tools of analysis by which scientific research can be developed as well as critiqued along feminist lines. I do not put forward these tools to create some special, esoteric "feminist" science, but rather to incorporate a critical awareness of gender into the basic training of young scientists and the workaday world of science.

Will Women Do Science Differently?

People often conflate the terms "women," "gender," "female," "feminine," and "feminist." These terms, however, have distinct meanings. A "woman" is a specific individual; "gender" denotes power relations be-tween the sexes and refers to men as much as to women; "female" des-ignates biological sex; "feminine" refers to idealized mannerisms and be-haviors of women in a particular time and place which might also be adopted by men; and "feminist" defines a political outlook or agenda.

The 1980s saw lively debates about the possibility of creating a "femi-nist science." If gender differences cut as deeply into the cultural fabric as historians and social theorists had discovered, so the argument ran, then the gender identity of the scientist must influence the content of science. These discussions were depoliticized in the 1990s and rephrased as a ques-tion: "Do women do science differently?" Even the prestigious journal *Science* jumped rather awkwardly onto the bandwagon with its query: "Is there a 'female style' in science?" Apparently not wanting to use the dreaded term "feminist," the journal's editors chose instead to focus on a "female style," implicitly attaching the question of scientific style to biological sex rather than to political outlook. The query bore a distant re-semblance to the question animating difference feminists: When women enter science, do they bring with them distinctive values and priorities? Strikingly, of the 200 women and 30 men who responded to the journal's follow-up survey, more than half said there was in fact a female style in science; only a quarter said there was not. The respondents were, of course, a highly self-selected group.[18]

The question of who or what might create change in science beneficial to women has been confused by Americans' mistrust of feminism. Femi-nism is for many still a dirty word, even among those who support the advancement of professional careers for women.[19] Especially within the sciences, people seem to prefer to discuss *women* rather than *feminism.* This refusal to acknowledge politics has led to a simple—and incorrect—equating of women entering the profession with change in science. Many women who enter science have no desire to rock the boat. Women who consider themselves "old boys" become the darlings of conservatives. (I once heard a well-established physicist refer to herself—apparently with-out irony—as an "old boy.") Institutions gain respectability by show-casing a few high-profile women while ensuring that fundamentals do not change. In some instances these "queen bees" even resist nurturing the progress of other women.

The reluctance to call a feminist spade a feminist spade has led many to overemphasize the importance of women as agents in the process of opening up science to women. In 1986 the physicist Mildred Dresselhaus, drawing on Rosabeth Kanter's elaborate study of corporate culture, of-fered the theory of "critical mass," suggesting that women experience fewer career obstacles once their numbers reach 10–15 percent of a par-ticular group. Small minorities tend to conform to dominant cultures, Dresselhaus maintained, but the presence of a slightly larger number of women can create an opportunity to reshape gender relations within a classroom, laboratory, department, or discipline.[20]

The critical-mass theory (with all its associations with nuclear fission) has been popular: in our highly gendered culture, many women do feel more comfortable with more women around. Certainly the commitment to equal opportunity requires attention to any remaining barriers imped-ing women's participation; our goal must be that their proportion in sci-ence equal their proportion in the larger population. Women, regardless of color or creed, should be represented equally in all aspects of life. Their full representation in the sciences will allow women the same freedoms men have long had to hold different outlooks and opinions and not to be considered *en bloc* as "the women" in a department or group.

Dresselhaus was concerned only with women entering and beginning to feel comfortable in the sciences. Others have attempted to study whether and how women approach science differently from men. The sociologist Gerhard Sonnert and the physicist Gerald Holton found in their study of 699 high-achieving scientists (men and women) that more than half thought women do science differently. The differences included being "inclined toward more comprehensive and synthetic work"; being more likely to avoid fields requiring head-to-head competition; being "more cautious and careful"; paying greater attention to detail; and choosing different subject areas for investigation. More women than men believed that gender played a role in their work as scientists; more men held the traditional view that science is and must remain gender-neutral.[21]

Other scholars have also focused on the presence of women as an important variable effecting change in what scientists study, or choice of research topic. Donna Holmes and Christine Hitchcock, surveying conference abstracts of the Animal Behavioral Society for 1981–1990, found that women disproportionately studied mammals, especially primates, while men tended to study fish, amphibians, and insects. Contrary to expectations, Holmes and Hitchcock did not find that women disproportionately studied females, though they were more likely than men to specify the sex of their subjects. Only within primatology did men and women tend to concentrate on animals of their own sex: women more often studied females or both sexes together, while men disproportionately studied only male primates.[22]

These studies and others like them tend to map political outlook onto sex, oversimplifying the process of democratizing science by making "women" alone the agents of that change. Linda Fedigan once remarked how dismayed she was when, after she spent many hours learning to identify individual female monkeys within a large group, many of her senior colleagues attributed her success to her sex; females are "empathetic," she was told, and this approach therefore is easy for them. In fact Fedigan's success depended on carefully implemented methods in primatology and long hours of observation.[23]

Saying that women's socialized qualities changed science overlooks the hard-won successes of twenty years of academic women's studies, the role of feminist men, and much else as well. Introducing new questions and directions into science (as into the social sciences or humanities) requires long years of training in a discipline, many years of attention to gender studies and feminist theory, universities and agencies that provide funding for that work, departments that recognize that work as tenurable, and so forth.

Because modern science is a product of hundreds of years of active shunning of women, the process of bringing women into science has required, and will continue to require, deep structural changes in the culture, methods, and content of science. Women should not be expected to succeed happily in an enterprise that at its origins was structured to exclude them. The assimilationist model of liberal feminism is inadequate. At the same time, the "difference feminist" model that suggests that women—because they have been socialized differently from men—carry the seeds of change with them into the laboratory is not rich enough. Some of the desire to attribute the successes of feminism directly to women derives from the fact that, historically, women as a group were excluded for no reason other than their sex. Some of the confusion derives from the fact that far more women than men have been feminists. Further confusion derives from the fact that nonfeminist women benefit from battles won by feminists.

The reason this question—Do women do science differently?—is so hotly disputed is that it remains in the realm of theory. The hypothesis that women will do science differently remains just that—a hypothesis in need of testing. (The same is true of its antithesis—that women will not do science differently). It is not obvious that gender has a stronger influence on science than do other political and cultural divides in North American society, such as class or ethnicity. To test such an idea, one would have to look at the kinds of perspectives that might be brought to science by African-American, Hispanic, Asian-American, Native-American, and Latina women (and so on), from upper-, middle-, and lower-class backgrounds, not to mention regional or other cultural differences. The lived experience of a woman from a Filipino immigrant family will be quite different from that of an African-American female graduate of Harvard, and different again from that of a white woman who grew up in rural Pennsylvania.

Testing the hypothesis that women *qua* women will do or have done science differently (or even that feminism would make a difference) would require a complex study of the history of science. Though both women and feminism are important variables, change in the methods and substance of science results from a vast array of subtle and not-so-subtle factors. No one would suggest, for example, that World War II was "caused" by Hitler or even by the rise of Nazism. Historians analyze long-term social and economic trends that caused instability in European and especially German society. They trace aspects of Nazism back to author-itarian aspects of Lutheranism, to weaknesses in the treaties structuring the peace after World War I, to the colonial backgrounds of European nations, to the frailty of democratic traditions in Germany, and so forth.

To understand changes in science that have been nurtured by feminism, we need to isolate and analyze the many factors involved. While the whole process of bringing women into science will no doubt have some impact on science, changes in the content of specific sciences such as medicine and primatology result from things like changes in gender ideologies and practices, receptive political climates, congressional action, and commit-ment to the advancement of women and their concerns (see Chapter 6). One would also have to look at other trends in the international structure of science, changes that have furthered goals often associated with femi-nism, but that may have nothing to do with women or feminism per se, such as the trend away from competition between single investigators to competition between internally cooperative groups.

After a while, change builds on change. The behavioral ecologist Judy Stamps has observed that much work from a "female point of view" is now being done by men who would not label themselves feminist but who borrow ideas on male-female relationships from the cultural climate in which they live.[24] A prominent male biologist told me that there are many more examples of feminist perspectives in biology than the ones I discuss in Chapter 8, but that since they have become part of mainstream biology, they are no longer identified as "feminist" or even associated in any way with women.

The complexity of the process of change that has resulted from the entrance of women into science does not mean we can relax policies de-signed to increase the number of women or scholarly attempts to under-stand gender dynamics in the content of science. Understanding the pro-cess of change can only enhance efforts to open science to women. (I would argue that the many attempts to increase the numbers of women in science through national and university programs that focus exclusively on women, rather than on institutions and ideologies, are not successful because they are based on impoverished understandings of the processes involved.) What is needed is a critical understanding of gender, of how it works in science and society.

Plan of the Book

In this book I evaluate current scholarship on gender and science in the United States, with occasional cross-cultural comparison. Gender studies of science grew up around the problem of how to increase the number of women working in science. Scholars tend to make a distinction between getting women into science and changing knowledge: getting women into science is generally considered the easier of the two. While career advancement for women is crucial, it is also clear that women will not become the equals of men unless certain aspects of science and scientific culture open up to gender analysis.

The book is divided into three parts: the first treats the history and sociology of women in science, the second treats gender in the cultures of science, and the third treats gender in the content of science. All three problems—getting more women into science, reforming the cultures of science, and opening new questions for research—depend on proper tools of gender analysis. All three are institutional *and* intellectual problems. One of my goals is to extract from current scholarship a set of useful analytical tools. These tools of gender analysis should be equally useful for advancing women's careers, restructuring laboratories, and overhauling research directions and priorities.

Part I (Chapters 1–3) provides a brief history of women in science and highlights how the culture of science, initially open to women, gradually closed as women were excluded from the grand party of humanity advertised in the Enlightenment proclamation that "all men are by nature equal." Science, and the medical sciences in particular, perpetrated studies of women's bodies that were used as evidence that women were not capable of taking on the duties of citizens in the state, participating in the professions, or producing works of intellectual depth and sophistication. I trace problems surrounding gender in science to their roots in the Scientific Revolution, the period often identified as foundational to the making of modern science.

Part I continues with a look at women scientists' employment opportunities today. Have the much-heralded U.S. intervention programs of the last two decades been successful in terms of employment for women?

How do the opportunities of women in U.S. science compare with those of women around the world? We may be impressed, for example, that (by one count) 47 percent of all physicists in Hungary are women—but are women successful there because physics has less of the prestige it enjoys elsewhere? I also track women's progress through the scientific "pipeline." The pipeline model—the idea that increasing the supply of girls interested in science will eventually result in greater numbers of women scientists—was found to be severely flawed in a study by the National Research Council.[25] Many universities persist, however, in employing this model for recruiting women into the sciences.

Part II (Chapters 4–5) concerns gender in the style of science. A major stumbling block along the road to women's equality has been the presumption that women should assimilate—that they should enter science on its terms, checking talents, traits, and styles not compatible with its cultures at the laboratory door. In 1834, in a paper wherein he also coined the term "scientist," William Whewell assured his readers that "notwithstanding all the dreams of theorists, there is a sex in minds."[26] Whether "sex" is located in the mind or in the culture (or neither or both) remains to this day a matter of some contention. Understanding the historical clash between the disparate cultures of science and of femininity is crucial to understanding the discomfort many women feel in the world of professional science.

In 1959 C. P. Snow identified two cultures, scientific and literary, between which loomed a gulf of "mutual incomprehension, . . . hostility and dislike, and most of all lack of understanding." Similarly, as we shall see, there exists a historically wrought clash between the cultures of science and women. Part of this conflict arises from the contest between professional and domestic life. In Chapter 5 I urge that domestic arrangements be considered part of the culture of science. The tension women (and increasingly men) encounter between family life and career is not entirely a private matter. Since the eighteenth century the celebrated "individual" has, in fact, been a male head of household. Professional culture has been structured to assume that a professional has a stay-at-home wife (today sometimes a husband) and access to her (or his) unpaid labor.[27]

Scholars have emphasized the consequences of exclusion for women, but what have been the consequences of women's exclusion for science and human knowledge more generally? In Part III I collect and analyze examples of gendering in scientific knowledge, exploring the question of how feminism has influenced the content of various sciences. Medicine provides one of the best examples of success for feminism. The National Institutes of Health Office of Research on Women's Health, founded in 1990, and the 1991 Women's Health Initiative represent major funding for neglected areas of women's health, such as osteoporosis and heart disease. But other sciences have had their successes as well. Paleoanthropologists and archaeologists have redefined "first tools" and, in the process, have reenvisioned women's role in human evolution. Primatologists, by taking females seriously as research subjects, have revised fundamental aspects of theories of sexual selection. And biologists, by challenging the attribution of human notions of masculinity and femininity to unsuspecting plants, animals, and even cells or bacteria, have revised and enlarged our understanding of human conception.

Gender analysis has made more progress in some scientific fields than in others. Effects of gender can be documented in the humanities, social sciences, and medical and life sciences, where research objects are sexed or easily imagined to have sex and gender. The physical sciences and engineering, however, have by and large resisted feminist analysis. This may be due to the extremely small number of people trained in both physics (or chemistry) and gender studies. Or is it due to the fact that the physical sciences are, as Steven Weinberg claims, as impersonal and free of human values as the rules of arithmetic?[28] These are the kinds of dilemmas we shall explore.

Most gender studies of science have focused on the United States and Western Europe. I offer a quick look into women's "indigenous knowledges" in other parts of the world. Feminists have expanded notions of what counts as science to include ways of understanding nature and responding to human needs not often regarded as "science." I focus here on women's traditions because they have generally been undervalued. Women—in Western countries and elsewhere—are generally considered recipients of knowledge rather than generators of knowledge. I draw attention to women's indigenous knowledges in the hope that this can be integrated more centrally into gender studies of science.

Terminology

I should say something about how I shall be using charged and often wobbly terms like "feminism," "gender," "sex," "women," "men," "male," "female," and "science." "Feminism" means very different things to different people; the many variants of feminism array themselves along a spectrum of philosophical and political outlooks. Notable feminists include men such as the seventeenth-century Cartesian François Poullain de la Barre, who declared that "the mind has no sex," and the nineteenth-century English liberal John Stuart Mill, who fought for women's rights. Feminism defines a perspective, not a sex. While historically most feminists have been women, associating the term too narrowly with women alienates sympathetic men and, more seriously, leaves unanalyzed how men have contributed to and been constrained by rigid notions of masculinity.

The term "gender" was introduced in the 1970s in efforts to check the then-rampant biological determinism, the point being to distinguish culturally specific forms of masculinity and femininity from biological "sex," construed as chromosomes, physiology, and anatomy. Biological determinists, then as now, grounded certain masculine traits, such as acute spatial relations, in male anatomy. The fashionability of the term "gender," however, resulted in its expropriation. Gender today is often used improperly as a PC code word for "sex," "woman," or "feminist." It is more properly used to refer to a system of signs and symbols denoting relations of power and hierarchy between the sexes. It can also refer to relations of power and modes of expression within same-sex relations.

Gender functions in different ways. Gender *ideologies* prescribe acceptable traits and behaviors for men and women. Europeans and Americans since at least the eighteenth century, for example, have been mesmerized by the notion of modest and delicate women protected by robust and valiant men. Gender ideologies are specific to region, religion, age, class, ethnicity, and so forth. Africans and many other non-Europeans have tended not to fit prevailing European and American visions of man- or womanhood. Gender *identity* denotes how any individual man or woman appropriates aspects of gender ideologies as part of his or her sense of self. Individuals' identities can change according to context, setting, and time. A woman may act "feminine" in a boardroom, say, but not among her close friends. Finally, gender *ascription* refers to behaviors expected of an individual because he or she is male or female. Confident women may be seen as aggressive because they transgress expectations of feminine comportment. "Gender," then, denotes multidimensional and changing understandings of what it means to be a man or a woman within particular social settings. It is historically contingent and constantly renegotiated in relation to cultural divides such as status, class, and ethnicity. While any particular man or woman may reject any particular set of gender attributes, he or she is nonetheless subject to the shifting rules and regulations of gender.

"Sex," by contrast, functions within gender studies to designate less malleable aspects of biology (though today there is an increasing appreciation of the mutability of sex, as for example when turtle and tortoise eggs incubated at 16–23 degrees Celsius produce males, while those incubated at 32 degrees or higher produce females). The term "sex" can have a variety of meanings. It can refer to highly ritualized romantic encounters; it can refer reductively to the exchange of genetic material between organisms (bacteria can have "sex" but probably not romance); it can refer to the biology of an individual ("male" meaning to produce gametes equivalent to sperm, and "female" meaning to produce eggs). Biologists tend to blur humanists' neat distinction between sex and gender by employing "gender" to refer to secondary sexual characteristics.

While scholars have struggled to hold onto the analytical distinctions between "sex" and "gender," there is a growing need to understand the relationship between these two concepts, as evidenced especially in work on the history of the body and in medicine and public health. Nancy Krieger and Sally Zierler suggest two complementary concepts to clarify the interdependent relation between biology and social expressions of gender. "Gendered expression of biology" refers to how biology influences gender—as, for example, when women's ability to become pregnant has been used to restrict their employment. "Biologic expression of gender" refers to how gender is imprinted directly onto the flesh-and-blood body, in ways that may not be associated with biological sex: bodies formed by cultural ideals of thinness, feet deformed by high heels, or, a hundred years ago, ribs broken by corsets.[29]

I will often talk about "women," the historical actors who, individually, have a sex and presentations of gender. Regardless of race, creed, sexual identity, or merit, all women—for no reason other than their sex—were prohibited from studying at European universities from the universities' founding in the eleventh century until the late nineteenth century. Similarly, all women, even grand property owners, were excluded from the rights of citizenship in the democracies of the Western world until the twentieth century. Women as a group have also been protected from exposure to lead and other occupational hazards, poison gas, Saddam Hussein's troops in the Persian Gulf, and until recently enemy fire at home and abroad. Women do sometimes have a common history. But they have also experienced history differently. Some women were slave owners, others were slaves; some women live below the poverty line, others work to cut welfare benefits; some are mathematically inclined, others have theatrical talents. It is sometimes appropriate to talk about women as a group, sometimes not.

There is also confusion surrounding the term "science." The project to investigate gender in science is not and should not be viewed as antiscientific. Nature, after all, is infinitely rich; there is much in nature we do not know. What we do know is influenced by our history and our values, our national and global priorities; sources of funding and patterns of patronage; the structure of academic institutions, markets, and information networks; personal and professional experiences; technologies and relations with foreign cultures; and much else besides. Culture does not construct reality, but works, as Evelyn Fox Keller has put it, "to focus our attention in particular ways, conceptually magnifying one set of similarities and differences while dwarfing or blurring others, guiding the construction of instruments that bring certain kinds of objects into view, and eclipsing others."[30] The goal of revealing gendered structure and polity in science extends the process of continuous critique that is part of the ordinary and remarkable workings of science.

1

WOMEN IN SCIENCE

We thought all you had to do was get more women into the pool—into graduate schools and tenure-track positions—and automatically they would move into the faculty and into industry, and so on. We were naive.

NEENA SCHWARTZ, neurobiologist, 1992

A woman who . . . engages in debates about the intricacies of mechanics, like the Marquise du Châtelet, might just as well have a beard; for that expresses in a more recognizable form the profundity for which she strives.

IMMANUEL KANT, philosopher, 1764

1

Hypatia's Heritage

I~N HIS~ lectures at the University of Uppsala in the 1740s, Carl Linnaeus taught that "God gave men beards for ornaments and to distinguish them from women."[1] The presence or absence of a beard not only drew a sharp line between men and women in the eighteenth century, it also served to differentiate the varieties of men. Women, black men (to a certain extent), and especially men of the Americas lacked that masculine "badge of honor"—the philosopher's beard. As Europe shifted from an estates society to a presumed democratic order, sexual characteristics took on new meaning in determining who would and who would not do science.

Recovering the accomplishments of great women scientists—from Hypatia, the renowned mathematician of ancient Greece, to Marie Curie—became a central task in the 1970s. Two challenges lent urgency to this project. The first was the need to find women who had indeed created science in order to counter the notion that women simply cannot do science, that something in the constitution of their brains or bodies impedes progress in this field. The second was the desire to create role models for young women entering science—"female Einsteins"—to counterbalance male stereotypes.

The question of women's place in science was not a new one. In 1405 Christine de Pizan, said to be the first woman to earn a living by her pen, asked if women had made original contributions in the arts and sciences:

> I realize that you are able to cite numerous and frequent cases of women learned in the sciences and the arts. But I would then ask you whether you know of any women who, through strength of emotion

and of subtlety of mind and comprehension, have themselves discovered any new arts and sciences which are necessary, good, and profitable, and which have hitherto not been discovered or known. For
it is not such a great feat of mastery to study and learn some field of
knowledge already discovered by someone else as it is to discover by
oneself some new and unknown thing.

De Pizan's fictional interlocutor, "Reason," gave the answer of the modern
historian of women: "Rest assured, dear friend, that many noteworthy
and great sciences and arts have been discovered through the understanding and subtlety of women, both in cognitive speculation, demonstrated
in writing, and in the arts, manifested in manual works of labor." Among
the inventions de Pizan attributed to women were the making of bread,
the dyeing of wool and making of tapestries, and the art of constructing
gardens and cultivating grains.[2]

Christine de Pizan's work was preceded and followed by a number
of encyclopedias of famous women. The first was Giovanni Boccaccio's
De mulieribus claris (1355–1359), featuring short biographies of 104
women, mostly queens (real and mythical) of the ancient world. The encyclopedia format—the most common type of history of women in science
from the fourteenth through the nineteenth centuries—was developed by
those who wished to argue for women's greater participation in science.
Encyclopedists collected names of distinguished women in order to prove
that women were capable of great achievements and should be admitted
to institutions of science. In 1690, for example, the French man of letters
Gilles Ménage published an encyclopedia of women accomplished in
ancient and modern philosophy as part of his proposal to admit women
to the Académie Française, the first academy in France's great system,
founded some thirty-one years before the prestigious Académie Royale
des Sciences.[3]

It was not until the late eighteenth century, however, that the first encyclopedia appeared devoted exclusively to the history of women's
achievements in the natural sciences. In 1786 the French astronomer
Jérôme de Lalande included in his *Astronomy for Ladies* the first short
history of women astronomers. In the 1830s the German physician Christian Friedrich Harless presented an evaluative history of the contributions
of women to all fields of natural science. As was popular in the heyday
of high Romanticism, he also argued that men and women display distinctive scientific styles: men search to uncover the causes underlying

appearances and to discover laws in life and nature; women search nature for expressions of love.[4]

The European women's movement of the 1880s–1920s sparked renewed interest in women's scientific abilities. In 1894, in Paris, the Saint-Simonians held the first conference on women and science, from which grew Alphonse Rebière's book *Les Femmes dans la science*. In the same year Elise Oelsner published her *Leistungen der deutschen Frau* (The Achievements of the German Woman), in which she paid close attention to women's scientific achievements. Both books followed the encyclopedia format, listing the women alphabetically, giving their names, dates of birth, the social conditions under which they had lived, their contributions and publications. Rebière included "professional scientists," as he called them, as well as amateurs and those patronesses whose contributions had aided "the progress of science." Appended to his work was a section of diverse opinions of famous people on the question "whether or not woman is capable of scientific pursuits."

By the late nineteenth century the encyclopedia approach was doomed, at least as a project for emancipation. Charles Darwin weighed in with his notion that genius is a virtual male monopoly: "If two lists were made of the most eminent men and women in poetry, painting, sculpture, music—comprising composition and performance, history, science, and philosophy, with half-a-dozen names under each subject, the two lists would not bear comparison." Antifeminists, such as Gino Loria in Italy, pointed out that even if one could recover enough distinguished women to fill three hundred pages, an equivalent project for men would run to thousands of pages. What woman, Loria trumpeted, can rival Pythagoras or Archimedes, Newton or Leibniz?[5]

In response, European and American feminists turned from the strategy of emphasizing the achievements of exceptional women and began to explore the barriers to women's participation in science. The first detailed work of this sort was published in America in 1913 by H. J. Mozans (a pseudonym for the Catholic priest John Augustine Zahm) under the title *Woman in Science*. Mozans's history was an impassioned attempt to show that whatever women had achieved in science had been through "defiance of that conventional code which compelled them to confine their activities to the ordinary duties of the household." He also provided a summary of discussions about women's capacity to do science, focusing largely on attempts by nineteenth-century craniologists to prove that the female brain was too small for scientific reasoning. Mozans urged women to join the scientific enterprise and unleash the energies of half of humanity; each woman was to act as a Beatrice to inspire her Dante, and in this way man and woman together would form a perfect "androgyn." Only then would the world enter a new Golden Age of "science and perfect womanhood."[6]

The works of Harless, Oelsner, Rebière, and Mozans are major landmarks in the history of women in science. Yet in their era the study of women in science was no more welcome than women scientists themselves within the academy. Nor was the picture to change with the emergence of the modern discipline of the history of science in the 1920s and 1930s. This new field, purporting to study the relation between science and society, did not consider the role of women in science. In the 1940s and 1950s those who did work on the history of women in science did so largely from outside the historical profession.[7]

In the 1970s, however, in the midst of a maturing women's movement and at a time when more and more feminists were assuming positions of power in both history and science, the study of the history of women in science took off. Women scientists contributed thoughtful autobiographies providing first-hand accounts of their struggle to make a mark on science.[8] Historians provided biographies of women scientists that deepened and broadened the work inherited from the nineteenth century.[9] These books call attention to exceptional women who defied convention to claim a prominent position in an essentially male world and also analyze the conditions that enhanced or diminished women's access to the means of scientific production. Without proper training and access to libraries, instruments, and networks of communication, it is difficult for anyone—man or woman—to make significant contributions to knowledge.

People tend to think that women have become scientists only in the twentieth century. Though today it would be difficult for anyone barred from university education or industrial laboratories to work in science, this was not the case in the seventeenth and eighteenth centuries. In this period, few men or women were full-time or salaried scientists. Some, like Galileo, were resident astronomers at princely courts; Bacon and Leibniz were government ministers, as well as men of letters. At the end of his life Descartes was in the pay of Queen Christina of Sweden as a tutor in natural philosophy and mathematics. This looser organization of science was one factor allowing women to find their way into scientific circles. It was not at all clear in this period that women were to be excluded from science.

⊙ Universities have not been good institutions for women. From their founding in the twelfth century until the late nineteenth and in some cases the early twentieth century, women were proscribed from study. A few women, however, did study and teach at universities beginning in the thirteenth century—primarily in Italy. They often flourished in fields, such as physics and mathematics, today thought especially resistant to women's incursion. The most exceptional example was the physicist Laura Bassi, who in 1732 became the second woman in Europe to receive a university degree (after the Venetian Elena Cornaro Piscopia in 1678) and the first to be awarded a university professorship. Celebrated for her work in mechanics, Bassi became a member of the Istituto delle Scienze in Bologna. She is rumored to have had twelve children (the historical record shows she had five), a burden that seems not to have interfered with her scientific productivity: each year she published the results of a new study—on fluids, the effects of air pressure, and the like. She also invented various devices for her experiments with electricity. The Englishman Charles Burney, who met Bassi during his tour of Italy, found her "though learned, and a genius, not at all masculine or assuming."[10]

The Milanese Maria Agnesi, celebrated for her textbook on differential and integral calculus (*Instituzioni analitiche,* 1748), was also offered a chair at the University of Bologna. She is often credited with formulating the *versiera,* the cubic curve that has come to be known as the "witch of Agnesi," though it had already been described by Pierre de Fermat. In trying to persuade her to take up a chair of mathematics and natural philosophy, Pope Benedict XIV proclaimed: "From ancient times, Bologna has extended public positions to persons of your sex. It would seem appropriate to continue this honorable tradition." She accepted this appointment only as an honorary one, and after her father's death in 1752 withdrew from the scientific world to devote herself to religious studies and to serving the poor and aged. By the 1750s the University of Bologna had offered a position to a third woman, Anna Morandi Manzolini, famous for her anatomical wax models showing the development of the fetus in the womb.[11]

The Italian model was not embraced across Europe. Germany experimented with the higher education of women, conferring two degrees (at Halle and Göttingen) in the eighteenth century; no degrees were awarded in France or England. Outside Italy no women were appointed professors, and within Italy the tradition of women professors did not continue. After about 1800 women were generally banned from institutions of higher learning. The mathematician Sofia Kovalevskaia was the next woman to become a professor in Europe; she was appointed to the University of Stockholm in 1889.

Why did Italy accommodate learned women in ways other European countries did not? The historian Paula Findlen has suggested that Bassi served to bolster Bologna's flagging patriciate, becoming a "symbol of scientific and cultural regeneration." With Bassi, the city could boast of a woman more learned than any other in Europe. The historian of science Beate Ceranski concurs that the traditions of Renaissance humanism, in which a woman could be admired for her learning, remained alive in the relatively small Italian city-states; but no woman—however great her learning—could hold such a position in the larger and more centralized states of France or England.[12]

Historians have traditionally focused on the decline of universities and the founding of scientific academies as a key step in the emergence of modern science. Except for a few Italian academies (the Istituto delle Scienze in Bologna, for example, and the Accademia de' Ricovrati), the new scientific societies, like the universities, were closed to women. The Royal Society of London, founded in the 1660s and the oldest permanent scientific academy, did not admit the eccentric but erudite Margaret Cavendish, Duchess of Newcastle, though she was well qualified (men above the rank of baron could become members without scientific qualifications). From its founding until 1945, the only permanent female member of the Royal Society was a skeleton in its anatomical collection.[13] The Académie Royale des Sciences in Paris, founded in 1666, refused to admit the prizewinning mathematician Sophie Germain (1776–1831); even the illustrious twentieth-century physicist Marie Curie was turned away. Nor did the Societas Regia Scientiarum in Berlin admit the well-known astronomer Maria Winkelmann (1670–1720), who worked at the academy observatory first with her husband and later with her son.

The prominence of universities and scientific academies today should not lead us to overemphasize their importance in the past. In the seventeenth and eighteenth centuries science was a young enterprise forging new institutions and norms. The exclusion of women was not a foregone conclusion. Several avenues into scientific work existed for women before the stringent formalization of science in the nineteenth century. As a result a number of women were trained and ready to take their place in the sciences.

In the early years of the scientific revolution women of high rank were encouraged to know something about science. Along with gentlemen *virtuosi,* gentlewomen peered at the heavens through telescopes, inspecting the moon and stars; they looked through microscopes, analyzing insects and tapeworms. If we are to believe Bernard de Fontenelle, secretary of the Académie Royale des Sciences, it was not unusual to see people in the street carrying dried anatomical preparations. Especially in Paris, wealthy women were ready consumers of scientific curiosities, collecting everything from conches, stalactites, and petrified wood to insects, fossils, and agates to make their natural history cabinets "the epitome of the universe."[14] In what I call noble networks—of natural philosophers, patrons, and illustrious consumers—well-born women often exchanged social prestige for access to scientific knowledge. The physicist Gabrielle-Emilie Le Tonnelier de Breteuil, marquise du Châtelet, for example, was able to insinuate herself informally into networks of scientific men by exchanging patronage for the attention of men of lesser rank but of significant intellectual stature.[15]

Royal women also formed crucial links across Europe as patrons of science. In 1650 Descartes was commissioned by the audacious Queen Christina of Sweden to draw up regulations for her scientific academy. Even the highest rank did not insulate women from reproach and ridicule. Many blamed Christina and the rigors of her philosophical schedule for Descartes's death, and for her philosophical prowess the queen was denounced as a hermaphrodite.[16]

Noble networks also flourished within salons, intellectual institutions organized and run by women. Like the French academies, salons created cohesion among elites, assimilating the rich and talented into the French aristocracy. Though these gatherings were primarily literary in character, science was fashionable at the salons of Madame Geoffrin, Madame Helvétius, and Madame Rochefoucauld; Madame Lavoisier received academicians at her home. There were, however, limits to this type of exchange. Just as privilege gave women only limited access to political power and the throne, nobility gave them only limited access to the world of learning. Because women were barred from the centers of scientific culture—the Royal Society of London, the Académie Royale des Sciences of Paris—their relationship to knowledge was inevitably mediated through men, whether those men were their husbands, companions, or tutors.[17]

Artisanal workshops served as another avenue into science for eighteenth-century women. The historian Edgar Zilsel was among the first to point to the importance of craft skills for the development of modern science. What Zilsel did not point out is that the new value attached to the traditional skills of the artisan also allowed for the participation of women in the sciences. Women were not newcomers to the workshop: it was in craft traditions that Christine de Pizan had located women's innovations in the arts and sciences—the spinning of wool, silk, linen, and "creating the general means of civilized existence."[18] In the workshop women's contributions (like men's) depended less on book learning and more on practical innovations in illustrating, calculating, or observing.

Whereas in France women's contributions to the sciences came consistently from women of the upper classes, in Germany some of the most interesting innovations came from craftswomen. The strength of artisans in Germany accounts for the remarkable fact that between 1650 and 1710 some 14 percent of all German astronomers were women—a higher percentage than in Germany today. Astronomy was not a guild, but the German astronomer of the early eighteenth century bore a close resemblance to the guild master or apprentice, and the craft organization of astronomy gave women a prominence in the field. Trained by their fathers and often observing with their husbands, women astronomers in this period worked primarily in family observatories—some in the attic of the family house, some across the roofs of adjoining houses, some on city walls.

In these astronomical families, the labor of husband and wife did not divide along modern lines: he was not fully professional, working in an observatory outside the home; she was not fully a housewife, confined to hearth and home. Nor were they independent professionals, each holding a chair of astronomy. Instead they worked as a team and on common problems. They took turns observing so that their observations followed night after night without interruption. At other times they observed together, dividing the work so that they could make observations that a single person could not make accurately. Guild traditions within science allowed women like the astronomer Maria Margaretha Winkelmann and the celebrated entomologist and botanist Maria Sibylla Merian to strengthen the empirical base of science.[19]

A number of other women of lower estate also contributed to science. Midwives, long before the recent enthusiasm for women's health initiatives, took charge of women's medicine (see Chapter 6). Wise women developed balms and cordials to prevent disease and cure ills. Outside Europe, women aided Europeans' forays into nature, preserving the health and well-being of foreign naturalists (mostly men) by preparing local foods and medicines. Women sometimes also served as local guides

for European expeditions; for example, much of the collecting and cataloguing for Garcia da Orta's *Coloquios dos simples e drogas . . . da India*
(1563) was done by a Konkani slave girl known only as Antonia. In one
instance an aristocratic woman, Lady Mary Wortley Montagu, served as
an international broker for women's knowledge. During her stay in Turkey as the wife of the British Ambassador at Constantinople, Montagu
learned of an old woman who—with her nutshell and needle—inoculated
children against smallpox. Though others had known of this practice,
Montagu was instrumental in introducing it into England.[20]

But women were not to be included as regular members of scientific
communities. In the nineteenth century the breakdown of the old order
(the guild system of artisanal production and aristocratic privilege) closed
to women what informal access to science they might have enjoyed. At a
time when the household was undergoing privatization, science was being
professionalized (a gradual process over several centuries). Astronomers,
for example, ceased working in family attic-observatories. With the increasing polarization of public and domestic spheres, the family moved
into the private domestic sphere, while science migrated to the public
sphere of industry and university.[21]

I place such emphasis on the Scientific Revolution of the seventeenth
and eighteenth centuries because it was at this time that the modern institutions and ideologies limiting women's participation in science were put
into place. Scientific institutions—universities, academies, and industries—
were structured upon the assumption that scientists would be men with
wives at home to care for them and their families.[22] The smooth working
of the professional world in many ways depended on the unacknowledged
contributions of wives who fed, clothed, and cared for their professional
husbands, providing well-run homes and ready support to further the
men's careers.

With the increasing professionalization of science, women wanting to
pursue scientific careers had two options. They could attempt to follow
the course of public instruction and certification through the universities,
like their male counterparts. These attempts, as we know, were not successful until the turn of the twentieth century. Or they could continue to
participate within the (now private) family sphere as increasingly invisible
assistants to scientific husbands or brothers. These talented women, including Margaret Huggins (wife of the British astronomer William Huggins), Edith Clements (wife of the ecologist Frederic Clements), and perhaps also Mileva Marič (wife of Albert Einstein), contributed quietly to

their husbands' careers, a phenomenon that persists even today. This became the normal pattern for women working in science in the nineteenth and into the twentieth century. Only occasionally did a woman, such as the X-ray crystallographer Kathleen Lonsdale, enjoy a husbandly assistant.[23]

Some wives, for example Marie Curie, did share scientific recognition with their husbands and go on to achieve fame in their own right. Marie and Pierre Curie were the first husband and wife to share a Nobel Prize (in 1903). But it was only after her husband's untimely death that Marie Curie assumed his professorship at the Sorbonne. This pattern of a wife taking her husband's place harks back to the guilds, and it is not a recognized route for professional development.

Historians of science have studied collaboration between husbands and wives and how, especially for women, marriage has served as an informal route into science. There is almost no scholarship on lesbians (the career advantages or disadvantages of this sexual identity) in science, nor on scientific collaboration between same-sex partners.[24]

In addition to working alongside their husbands, women performed what the historian Margaret Rossiter has called women's work in science. Numbers of anonymous women served as invisible crews and technicians supporting the man at center stage. They did various repetitive and tedious jobs, at times poring over astronomical star plates or cataloguing natural history collections, measuring tracks on bubble-chamber film, or calculating equations before the advent of computers.[25]

Women embarked on modern careers in science only after the women's movement of the 1870s and 1880s propelled them into universities. As they gradually gained admittance to graduate schools—by the twentieth century a prerequisite for serious work in science—they poured into Ph.D. programs. By the 1920s their numbers were at a historic high in the United States, with women earning 14 percent of doctorates in the physical and biological sciences. Between 1930 and 1960, however, the proportion of women Ph.D.'s plunged as a result of the rise of fascism in Europe, the Cold War, and McCarthyism in the United States. Women did not regain their 1920s levels of participation in academic science until the 1970s. A similar pattern characterized women in medical schools: their numbers peaked around 1910 only to fall again.[26]

Women made some gains in academic ranks during World War II: they held 12 percent of teaching positions in 1942 but 40 percent in 1946. But after the war, in what Margaret Rossiter has called the remasculinization of science, the "old girls" were moved aside. Universities seeking to increase their prestige raised salaries, reduced teaching loads, hired more Ph.D.'s, and restored faculty positions to men. One university president is quoted as having said, "We do not want to bring in more [women] if we can get men."[27] Rossiter has pointed out that even home economics, a science created and traditionally populated by women, suffered acute masculinization in this period.

Women's lot was worsened by the G.I. Bill, which provided qualified veterans with lavish benefits including five years of full tuition and a living allowance. Of the nearly 8 million veterans flooding American universities after the war, only 400,000 were women. While the number of men annually receiving science doctorates rose from 800 to 4,000 from 1946 to 1960, the number of women was held below 500. Women for the most part missed out on the "golden age" of postwar American science, a time of record growth in terms of monies spent, persons trained, and jobs created.[28]

Beginning in the 1960s and 1970s a number of factors conspired to encourage women to enter science. In 1964 Title VII of the Civil Rights Act (later strengthened by the Equal Employment Opportunity Act of 1972) outlawed discrimination on the basis of sex in education and employment. The time had passed when the chairman of Cornell University's biochemistry department could reject a qualified applicant simply because she was a woman. The launching of Sputnik in 1957 prompted a frenzy of recruitment, fueled by the sense that the United States needed more scientists to retain its competitive edge. In this atmosphere even women and minorities figured as valuable national resources. This, coupled with the renewed women's movement of the 1970s, brought about a boom in women's participation in science—a boom intensified by government funding of programs designed to attract more minorities and women into science and engineering. By 1995, 23 percent of U.S. scientists and engineers were women.[29]

The early history of women in science teaches us several things. First, it teaches that scientific institutions have taken many forms over the centuries and that the structure of these institutions can encourage or discourage women's participation. Second, it reveals that in modern industrial societies the division of labor between work and home remains a barrier to women's entering the professions. Third, history teaches that women's success in science depends on a variety of interdependent factors: the prestige of scientific institutions, the fortunes of war and peace, the political climate, the structure of the family vis-à-vis the economy. Many of the problems women face in science today—domestic versus professional responsibilities, the tenure clock versus the biological clock—have deep historical roots. And fourth, history dispels the myth of inevitable progress in respect to women in science. There is a sense that nature takes its course—that, given time, things right themselves. The history of women in science, however, has not been characterized by a march of progress but by cycles of advancement and retrenchment. Women's situation has changed along with social conditions and climates of opinion.

《② 2

Meters of Equity

W HAT IS the situation for women in U.S. science today? Collection of statistics on women in science began in earnest in the 1970s as part of the project to increase their numbers. Since 1982 the National Science Foundation has produced booklets entitled variously *Women and Minorities in Science and Engineering* and *Women, Minorities, and Persons with Disabilities in Science and Engineering*. It has become common for books and lectures about women in science to begin with statistical surveys.

Why statistics? Measuring discrimination does not remove it. But numbers enjoy the cachet of truth in our society, and statistics are thought to provide an objective measure of the status of women. Initially they were employed to prove just how disadvantaged women were in science. Now they more often serve to chart positive changes in women's employment opportunities and salaries.[1]

In the early 1980s Margaret Rossiter offered two concepts for understanding the mass of statistics on women in science and the disadvantages women continued to suffer. The first she called hierarchical segregation, the well-known phenomenon that as one moves up the ladder of power and prestige fewer female faces are to be seen (Figure 1). This notion is perhaps more useful than that of the glass ceiling—the supposedly invisible barrier that keeps women from rising to the top—because the notion of hierarchical disparities draws attention to the multiple stages at which women drop off as they attempt to climb academic or industrial ladders. Women now earn 54 percent of all bachelor's degrees in the United States (parity was achieved in 1982) and 50 percent of those in science. Women begin to drop off at the graduate level, where they earn 40 percent of all

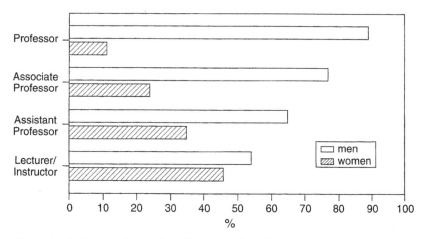

Figure 1. Ph.D. scientists and engineers employed in universities and four-year colleges, 1995. The representation of women in faculties of science and engineering falls off as one progresses through the ranks. Source: National Science Foundation, Characteristics of Doctoral Scientists.

doctorates (31 percent in science and engineering). Another drop-off occurs at the faculty level: in 1995 11 percent of full professors in all fields of science and engineering were women. Only three women were deans at the 311 accredited engineering colleges in the United States. That is less than one percent.[2]

Rossiter also discussed "territorial segregation" or how women cluster in scientific disciplines (see Appendix). The most striking example of occupational territoriality used to be that women stayed at home and men went out to work. Today, with women making up nearly half of the civilian labor force, that is no longer true. Women still tend, however, to be concentrated in low-paid occupations: 60 percent of white professional women are nurses, daycare workers, or schoolteachers, while nearly half of all African-American women in the labor force work as chambermaids, welfare service aides, cleaners, or nurses' aides.[3]

Territoriality also defines life for women in the academy. We all know that women are more likely to teach and do research in the humanities and social sciences than in the natural sciences and engineering. (Exceptions do exist; in 1994, for example, women earned 41 percent of biology Ph.D.'s but only 37 percent of history Ph.D.'s.) Territorial disparities are found within the sciences: in the 1920s and 1930s the three big scientific fields for men were chemistry, medical sciences, and engineering, while those for women were botany, zoology, and psychology—fields with less prestige and less money. Today women are concentrated in what are known as the "soft" sciences: the life and behavioral sciences and the social sciences, where salaries are relatively low regardless of sex (Figure 2). Few women are found in the "hard" or physical sciences, where prestige and pay are high. This may explain why only 9 percent of U.S. physicists are women: until the end of the Cold War physics was arguably the most prestigious field in American science.[4]

Women may well be tracked into certain specialties, such as pediatrics or gynecology in medicine; and the "feminization" of certain fields, such as women's studies, may endanger their funding and status. It is also the case that women cluster in certain fields because they feel comfortable there and are able to become leaders. One often hears that for certain academic positions good women "cannot be found," especially at the senior level. Perhaps departments are not defining positions in areas that have traditionally attracted women.[5]

Men fare better in traditionally female fields, such as nursing, than women do in traditionally male fields, such as physics or engineering. In 1991 women earned the overwhelming majority of Ph.D.'s in nursing (91 percent), and yet men held 4 percent of full professorships of nursing. In no field of science where women earn less than 10 percent of the Ph.D.'s do they hold 4 percent of full professorships. In 1992 women earned 9 percent of the Ph.D.'s awarded in engineering but made up only 1 percent of full professors. Even in psychology, a field in which women earn a majority of the Ph.D.'s (62 percent in 1994), they hold a far lower percentage of tenured positions (19 percent in 1994).[6]

In addition to hierarchical and territorial discrimination, women also suffer institutional segregation. While women now study at prestigious universities at about the same rate as men, they are rarely invited to join the faculty at top universities. Harvard University tenured its first woman in chemistry (Cynthia Friend) in 1989 and its first woman in physics (Melissa Franklin) in 1992. The sociologist Harriet Zuckerman has observed that "the more prestigious the institution, the longer women wait to be promoted." Men, generally speaking, face no such trade-off.[7]

Women's lower status in the scientific community is also reflected in their salaries. In 1993 the median salaries of women scientists and engineers holding doctoral degrees were 20 percent less than men's. This is true partly because women cluster in low-paying fields. But even within the same field women typically are paid 15–17 percent less than men.

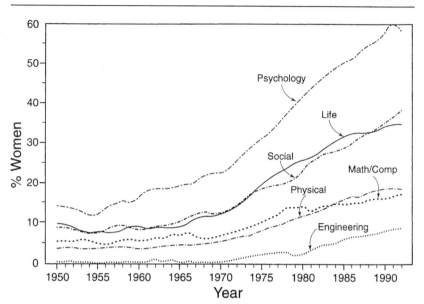

Figure 2. Percentage of Ph.D.'s in scientific fields awarded to women, 1950–1992 (three-year averages). Fewer women cluster in the physical sciences. Source: Vetter, *Professional Women and Minorities.*

There are also some salary differences among women that are worthy of note. African-American women scientists with 10–14 years' experience earn 3.4 percent less than European-American women scientists with similar qualifications. Among engineers, Asian women tend to outearn other women.[8]

More generally, trends show women earning less than their male colleagues whether employed in industry, government, or academia—though the wage gap in academia is the greatest. Only in engineering and chemistry do women enjoy higher starting salaries than men: for engineering, about $4,000 higher. After five years, however, these women engineers' salaries dip approximately $2,000 below men's and continue at a lower level.[9] There is, of course, nothing unusual about science in this regard: within the 800 largest U.S. companies, only 19 of the 4,000 highest-paid officers are women.

Some attribute women's lack of professional success (measured in terms of rank and salary) to their lack of experience. As a group, women scientists are younger and less experienced than men scientists (in 1993 the average employed woman scientist or engineer had held her doctorate for approximately ten years, while the average man had held his for sixteen years). The National Science Foundation has found, however, that after correcting for age, experience, and education, discrimination remains the only explanation for women's and minorities' poor positions and salaries. Margrete Klein, the director of NSF's Women's Programs, reports that women tend to leave science and engineering in their late thirties, just as they should be maturing into positions of leadership. Women, she says, too often become disillusioned. They may choose to leave science rather than fight an uphill battle for recognition and rewards.[10]

Positive signs are on the horizon. In 1996 salaries for women in professional fields increased to 85–95 percent of men with similar jobs. Younger women in the United States (childless women between the ages of 27 and 33) earned nearly the same (98 percent) as men in their age group. In the workforce as a whole, women make 74 percent of what men make (up from 59 percent in the 1970s).[11]

Studies of women in science, more than other areas of women's studies, have focused on European-American women. What is currently known about the situation for minorities in U.S. science? Many of the problems that beset women, who constitute a majority (51 percent) of the population, also plague minorities. Most are represented in science far below their numbers in the working population. African Americans make up 11 percent of the workforce but only 3 percent of employed scientists and engineers. They are poignantly absent from science faculties: of the 60,347 full-time natural science faculty members in the United States in 1987, only 1 percent were African American. In the early 1990s the University of Chicago had only 21 African-American faculty members in all fields among a faculty of 1,226. Hispanics are also poorly represented in science, making up 8 percent of all workers but only 3 percent of scientists and engineers. Native Americans so rarely become mainstream scientists or engineers that their participation cannot be measured statistically. Only Asian Americans are "overrepresented" in science, making up 9 percent of all scientists and engineers working in the United States in 1993 but only 3 percent of the population. One reason for documenting their overrepresentation has been to determine whether or not Asian Americans should qualify for affirmative-action programs (in most instances they do not, and at some universities upper limits have been set on their enrollment). Despite their general academic excellence, Asian Americans often encounter difficulties advancing. Many remain in the rank and file of bench scientists.[12]

In the United States "minority" has often meant men (and specifically African-American men) and "women" has meant whites. As expressed in a book on contemporary black women's studies, "all the women are white, all the blacks are men." The National Science Foundation has only recently begun to break down statistics on sex by ethnicity and those on minorities by sex. The NSF persists in its astonishing message that minorities are better represented among female than among male scientists. In 1990, for instance, 11 percent of women scientists and engineers were African American, compared with only 7 percent of men scientists and engineers. In absolute numbers, however, male African Americans in science outnumber female African Americans by about two to one. African-American women scientists are more likely to be employed (33.5 percent of Ph.D. holders) than European-American women scientists (21.5 percent).[13]

Minorities (undifferentiated by sex) have experienced an increasing wage gap. While African-American and European-American scientists earned about the same in 1972, a decade later African Americans' salaries were 6 percent lower than those of European Americans. In 1991 Ph.D. scientists of color earned about 9 percent less than their European-American counterparts. The biggest gap was in chemistry, where European Americans earned 22 percent more than African Americans.[14]

It is important to recall just how recently blatant discrimination against African Americans was considered respectable in the United States. In 1967 miscegenation was still banned by law in a number of states. In 1962 no African American of either sex was permitted to study for a Ph.D. in Virginia, though the state paid full tuition for African Americans to go to college elsewhere. Minority women often encounter the double bind of racism and sexism. Vivienne Malone Mayes's experience at the University of Texas is a case in point. In 1962 Mayes became the third black woman in the United States to earn a Ph.D. in mathematics (the first two were awarded in 1949). She found that because of her race she was ineligible for a teaching assistantship and banned from some classrooms. Her race also barred her from the café where her adviser and classmates met for informal discussions. Only after winning the fight to desegregate the café did she discover that women, whatever their race, were not welcome. Reflecting on her experiences some years later, she wrote: "I was the only Black and the only woman . . . my isolation was complete."[15]

Cross-Cultural Comparisons

How does the United States compare to the rest of the world with respect to women's participation in science? This is a difficult question to answer; cross-cultural information is only now being gathered. Among the industrial nations few significant differences in overall patterns are evident, but there are some interesting variations. In 1994 Jim Megaw released statistics revealing that some of the most "advanced" nations have the lowest proportions of women among their professors of physics. The United States, with 5 percent, ranked seventh among 29 countries; only Japan, Canada, Germany, Switzerland, Norway, and Korea were the same or worse. Countries such as Italy, the former Soviet Union, and Portugal did much better, with over a quarter of their faculty positions in physics held by women. These differences remain unexplained. Preliminary research suggests that in many of the countries where women do well in science, math and science are mandatory in high school. Women also do better in countries where children attend single-sex schools.[16]

Differing attitudes toward work and family also help explain cross-cultural differences in women's success in science. Some have proposed that women do better in predominantly Catholic countries (Italy and France), where the extended family still provides an intimate support network for childrearing. The absence of the Protestant work ethic further allows for more flexibility in the work place. This argument, however, ignores the need for job mobility; couples tied to extended families might not be able to move easily in order to advance their careers.

This argument also overlooks the role of national governments in establishing social supports for working parents. The example of the two former Germanies shows that political policies are at least as important as religious traditions. The Federal Republic of Germany has the distinction of being one of the worst possible places for academic women. Women's representation among senior faculty in five sciences—biology, physics, chemistry, math, and the geosciences—stands at a mere 2 percent. At the elite Max Planck Institutes in 1997, only 2 percent of the scientific members were women.[17] This contrasts with the former German Democratic Republic, where women held a significantly higher proportion of academic positions overall and constituted 9 percent of all physicists. These numbers have dropped dramatically since reunification in 1989; the current German government has admitted that women scientists from the former East have been adversely affected by unification.[18]

How do we explain these differences in countries that until World War II shared many religious and cultural traditions? After the war West Germans continued to support the notion that women should devote their lives to *Kinder, Küche, and Kirche* (children, kitchen, and church). The Soviet-dominated East Germany, in contrast, encouraged women's participation in the workforce and created the national daycare facilities. These facilities allowed families to balance careers and child care, but it was also clear that the balancing act was women's work. In East Germany, as in the former Soviet Union and currently in the United States, it has been women who shoulder the double burden of domestic labor and full-time jobs.

Social supports for working parents do not always guarantee women's success in science. Swedes enjoy government-subsidized daycare; Sweden is one of the few countries I have visited in recent years where the women I talked to did not seem anxious about domestic arrangements. Yet only 6 percent of university professors there are women. What it means to be a professor in Sweden and in Europe more generally, however, is different from what it means in the United States: university systems are more elite and "professors" are closer to the American category of distinguished or chaired professors.

Are women doing better in what used to be called the Third World? When talking about women in Third World science, people tend to consider women's participation primarily in university-research science. Scientific institutions in these countries are generally modeled on those in the United States or Europe; consequently, patterns of women's opportunities are similar to those in the United States or Europe. Transfer of technology often includes the inadvertent transfer of European and American gender ideologies and divisions of labor. Men are more often encouraged to go into "manly" subjects like physics, chemistry, mathematics, and engineering. When schooled at all, women are commonly educated to be nurses, secretaries, or teachers. In Ghana women constitute 9 percent of natural scientists, 4 percent of engineers, and 13 percent of social scientists—a pattern recognizable to American and European eyes.[19]

In some cases, however, women's participation outstrips that in the United States and Europe. In China the proportion of women in science and engineering is higher than in the United States (32 versus 16 percent in 1988). Women make up 17 percent of academic scientists (though girls had to score 5 percent higher on entrance exams than boys to be admitted to universities). Even in China the proportion of women decreases with the prestige of the position: the Chinese Academy of Sciences has only 3.5 percent women members.[20]

Women scientists and engineers also flourish in Turkey. Today 32 percent of natural scientists (down from 44 percent in 1946), 30 percent of medical personnel, and 24 percent of engineers are women. After World War I Turkish leaders made improving the status of women part of their plan for modernization. Elite women, who could hire housekeepers and nannies, responded favorably to the new opportunities in the natural sciences. In this period of rapid modernization the advantages of elite social standing often outweighed the disadvantages of sex. To fill university posts, governmental authorities preferred women from the upper classes to men of lower social standing. (Even today only 1 percent of Turkish women and 2 percent of Turkish men attend university at all.) Though women were well represented in the natural sciences, they were conspicuously absent from the faculties of law and political science—fields more closely associated with power and privilege in Turkey.[21] Men are often discouraged by family members from pursuing low-paying careers in academia. Women, who are not necessarily expected to support their families, are better able to pursue sometimes poorly paid careers in research science.

Women's Indigenous Knowledges

Can Western tools of gender analysis be extended to other cultures? In many instances Western feminism is embraced by women around the world; in others feminism is no more welcome in what has been called the Third World than are Western sciences. Feminism's emphasis on women's equality is often seen as just another Western value being foisted onto cultures with their own traditions.[22]

One aspect of women's participation in science that is just beginning to be studied is their involvement in what is often called "indigenous" science traditions. Many precolonial knowledge systems have been destroyed; remnants of many remain or are being revived. One might hypothesize that, just as women played an important role in midwifery in Europe and America (see Chapter 6), women in other parts of the world might be better represented in indigenous—or nonprofessionalized—

knowledge systems. In the early 1990s the United Nations convened a working group to study women's indigenous knowledges to encourage research in this area.[23]

The term "indigenous knowledges" is admittedly problematic, but I use it here to refer to knowledges not recognized as "science." Achoka Awori has suggested that this term—meaning knowledge systems native to a place—is preferable to the term "traditional science," which has been used to refer to rote application of existing technologies. Yet another term, "ethnoscience," has been used primarily by anthropologists to refer to the knowledge systems unique to particular cultures. Because anthropologists commonly study "primitive" peoples, ethnoscience has tended to be devalued by Westerners as primitive—static, backward, based on myth and superstition. Furthermore, the philosopher of science Sandra Harding has argued that "ethnoscience" applies as much to Western as to other forms of science. "Maximizing cultural neutrality," she maintains, "is itself a culturally specific value." Abstractness and formality express "distinctive cultural features, not the absence of any culture at all."[24]

Though some still object that "indigenous" is too closely bound to the notion of native or underdeveloped, I will employ the now widely used term "indigenous science" to refer to science traditions that do not fit the research-university model. These knowledges exist also in industrialized countries, sometimes alongside or in contest with U.S. or European science. They are sometimes referred to as "local knowledges." In fact, however, as in the case of midwifery, many of them are not local but disseminated widely from culture to culture.

Examples of women's indigenous knowledges have clustered around agriculture and forest management because in many cultures women are in charge of food and food preparation. The mathematician Ram Mahalingam has suggested that Tamil women in India have actively developed certain areas of mathematics crucial to *kolam,* rice flower drawings, and *pallanguzhi,* a bead game; this has yet to be fully investigated.[25]

The physicist Vandana Shiva has noted several examples of women's indigenous knowledges emerging from their work breeding plants and animals to meet the nutritional and medical needs of their families in India. One example is women's forest management techniques, where what is carried away from the forest in the form of fodder, water, fuel, and fibers is returned to it as compost in efforts to create sustainable ecosystems. In this area women have developed a technique called "lopping," the selective thinning of leaves from oak trees. Oak leaves, along with a mixture of dried grasses and agricultural byproducts, are fed to cattle through the late autumn, winter, and into spring. Lopping keeps the leaves soft and palatable for the cattle in the summer months. It allows the trees to be used for fodder while it simultaneously enhances their density and productivity.[26]

Another example of women's indigenous knowledges comes from the Andes, where for centuries Quechua women have bred and preserved potatoes and their seeds. We can better understand the achievement of these *semilleras* or "keepers of the seeds" when we consider that a mature ear of wild maize was about an inch long and as thick as a pencil; wild potatoes are equally unrecognizable by modern standards.[27] Present-day seed custodians are repositories of agronomic and physiological knowledge concerning Andean roots and tubers. They meet annually to exchange produce, find new seeds, and share knowledge about production, conservation, and use. An experienced *semillera* can distinguish dozens of varieties of potatoes and knows about their time of maturation, yield, disease susceptibility, cooking properties, perishability, and the like. One woman may manage up to fifty-six varieties of potatoes and other tubers.

Andean women typically select and classify their potatoes according to four criteria: type of cultivation, edibility, processing required, and resistance to frost and pests. What we might call "varieties" are further classified by skin and meal color, meal consistency, shape of the tuber, and depth and configuration of the eyes. "Subvarieties" are sorted primarily by tuber color. A woman cultivates diverse potatoes both to provide a balanced diet for her family and to conserve soil fertility. She also pays attention to her family's palates. The tastiest potatoes, often raised in special plots, are reserved for festival days.

Since 1950 women have also distinguished between native potatoes (called "gift potatoes" or "colored potatoes") and those introduced by the Peruvian National Potato Program (called "improved potatoes" or "white potatoes"—for their white meal). Improved varieties are bred for their yield (two to three times that of native potatoes) and marketability in urban areas. They do not, however, store well, or produce viable seeds; and they require the use of chemical fertilizers, insecticides, and fungicides. Andean women judge the "improved" potatoes inferior to the native ones and rarely feed them to their own families.[28]

There are interesting gender dynamics in discussions concerning the preservation of this knowledge. In Peru there are two models for preservation. The first revolves around the International Center of the Potato

(Lima), where specimens are preserved *ex situ* in one centralized laboratory. This shift from countryside to laboratory brings a shift in personnel:
women who are central to cultivating potatoes in the countryside are
marginal to preserving the potatoes in the Western-style laboratories. An
alternate model for preservation requires preservation *in situ*, where the
seeds continue to adapt to very particular climatic and cultural conditions.
This model requires that the project of preserving biodiversity be tied to
that of preserving cultural diversity. In order to preserve their seeds *in
situ*, Andean *semilleras* must be able to control the use of their lands
within a global economy that allows them to make a living and feed their
families.[29]

There is nothing sacred or mystical about the fact that these particular
knowledges of nature have been developed by women. Women's work in
seed preservation emerges from the sexual division of labor, property, and
power in particular cultures.[30] The same kind of work could be done by
men under different conditions.

We sorely need more cross-cultural comparison of the diverse factors—
social, economic, institutional, cultural, and political—encouraging and
discouraging women's participation in science. The few studies that have
been done suggest that, across cultures, the number of women in a particular science tends to be inversely proportional to its prestige: the higher
one goes in the scientific hierarchy the fewer women one finds. Where the
statistics look good, the jobs may not be highly valued and may therefore
be open to women. In Venezuela, where women make up 54 percent of
the medical personnel, as late as 1991 not one had become a member of
the National Academy of Medicine.[31] Where science counts most in terms
of international prestige, we seem to find the fewest women.

Publication Counts

Statistics are just one way scholars have attempted to objectify the sometimes subjective experience of discrimination. Publication and citation
counts are another. Are these useful tools of gender analysis? Publication
counts are important because they are very often used to determine tenure
decisions, increasingly even within the humanities. But what do they actually tell us?

Publication counts, which emerged in the 1960s to measure scientific
productivity, have been employed to evaluate charges of overt and covert
discrimination. In 1979 the National Research Council blamed the low

numbers of women in science on the "sex discrimination practiced for many years in some graduate science departments."[32] Jonathan Cole countered by arguing that neither overt nor covert discrimination accounted for the dismal condition of women in science. In his book *Fair Science* (1979) he held that the failure of women to rise to the top results from their lesser contribution to scientific knowledge, their lower productivity, lesser citation rates, self-deletion from the competitive process, and so forth. According to Cole, the perception of discrimination—the many stories women tell—is not borne out empirically. Science is "fair"; women are to blame for their poor showing. While statistics have been gathered, more often than not, to suggest that there is discrimination in the sciences, publication counts have been used to demonstrate that women are, in fact, reaping rewards equal to their merit as active scientists.

Cole devised elaborate publication counts to show that women are indeed rewarded in proportion to what they produce. In his view people too often conflate the simple fact of women's underrepresentation in science with discrimination. Women constitute only 5 percent of the membership of the National Academy of Sciences, for example, because they have not produced the science to put them there. Cole went so far as to claim that, when productivity is taken into account, there is "a slight tendency for women to be overrepresented in higher-quality departments," at places like Harvard, Berkeley, Stanford, or Princeton.[33] He argued further that affirmative action programs had eliminated any residual discrimination against women and that continuation of these programs would result in "reverse discrimination"—women being rewarded beyond their merit. Cole's claims spawned a minor industry among sociologists aimed at precision in measuring scientific productivity. This literature has focused narrowly on the individual scientist—his or her life choices and how they influence a career.

In 1984 Cole joined forces with Harriet Zuckerman to publish "The Productivity Puzzle," in which they argued that the difference between the sexes in research output is significant and also "puzzling." They showed that for a cohort of scientists receiving Ph.D.'s in 1970 women's publication rate was approximately half of men's across all fields of science. Studying matched pairs (men and women who received Ph.D.'s in the same departments in the same years), Cole and Zuckerman found that gender differentials in publication started early in a career and grew as a scientist matured. These differentials were reduced, but not eliminated, when rank and type of institution were held constant. Women's poor showing resulted from the significant differences at the high and low ends of productivity: men at prestigious institutions far outproduced anyone else, publishing five or more papers per year. Twelve years after their Ph.D.'s, 22 percent of the women (compared with 11 percent of the men) had not published a scientific article.[34]

One should keep in mind that productivity studies look at relatively old data sets: Cole and Zuckerman's paper was based on a cohort of scientists who received their Ph.D.'s in 1970; J. Scott Long's work published in 1992 was based on scientists who received their Ph.D.'s between 1950 and 1967. Considering the rapid gains women have made in the past two decades, one might expect that their productivity would have improved dramatically. In 1984, however, Cole and Zuckerman concluded that social change wrought by the women's movement and affirmative action had done nothing to change women's overall productivity.[35] While women increased their numbers among the most prolific scientists (from 8 percent for women getting their Ph.D.'s. in 1957–58 to 26 percent for women getting theirs in 1970), they were still underrepresented in this group. It will be crucial in this regard to see what studies of women receiving their degrees in the 1980s and 1990s—a time of rapid institutional reform for women—will reveal.

Cole and Zuckerman found few significant impediments to women's productivity. Women encountered few difficulties in publishing. They also encountered few problems in their collaborations. Women collaborated with colleagues just about as often as men; over the course of their careers both men and women worked with between 2.5 and 3.1 colleagues per paper. Other sociologists have pointed out, however, that though women collaborate as often as men, women work with fewer people, thus narrowing their information networks. Simply measuring the frequency of collaborations ignores the fact that women are more likely than men to collaborate with a spouse (6–10 percent of women compared with 1–2 percent of men). This difference may result from the fact that women scientists marry other scientists at a very high rate.[36]

Women working with men, and especially with their husbands, have traditionally been considered the lesser partners. It is commonly assumed that Pierre Curie was mainly responsible for the Nobel Prize he shared with his wife, Marie, but in fact she shared equally in the work. It has been said that Enrico Fermi "gave" his colleague Maria Goeppert Mayer the model for her Nobel Prize–winning work in physics. Some women say they shy away from working with men for fear of becoming entangled in rumors about sexual encounters. For different reasons, few women collaborate with other women. The paleobiologist Anna Behrensmeyer, celebrated as the mother of taphonomy, found that when she regularly wrote with another woman "there were snide remarks about how taphonomy was being done only by women." Few would make similar remarks about long-term collaborations between men.[37]

Amazingly enough, Cole and Zuckerman also ruled out marriage and the family responsibilities it entails as a significant factor in women's allegedly poor research performance. They found that marriage actually increased women's research productivity, despite the fact that marriage dramatically decreased women's ability to change jobs in order to advance. Even more amazingly, Cole, Zuckerman, and others emphasized that women with two or fewer children are just as productive as those without children.[38]

J. Scott Long, however, revealed that the benefits of marriage for a woman's productivity may have to do not with her domestic situation but with the fact that women's mentors (87 percent of whom are men) feel more comfortable collaborating with married women. Long found that being married doubles the odds that a female postdoc will collaborate with her adviser. Long also found that having children decreases a woman's chances of collaborating with a male mentor: a mother has less time to spend at the lab and less flexibility in her hours; and, until recently, she may not have been considered a serious scientist. Consequently, Long argued, it is the lack of opportunities for collaboration, not the direct effect of having young children, that decreases these women's productivity. For men, by contrast, family issues are rarely significant. Neither marriage nor children has much effect on their productivity or their relationships with mentors (though it may in the future as more women become mentors). Long also noted that women's mentors are less productive (publishing 25 percent fewer articles than the average man's mentor), less prestigious, and more often female.[39]

What does counting publications really tell us? Cole and Zuckerman urge that, like it or not, research performance is the coin of the realm when it comes to garnering resources and rewards. Their assumption that rewards blindly follow merit, however, flies in the face of Long's 1978 finding that productivity does not determine job placement. A recent Swedish study showed that women need to publish three times more papers to rank equally with male colleagues. Whatever the outcome of the debate about scientific productivity, the majority of researchers in this area (including Zuckerman) agree that, other things being equal, men and women with equivalent research records do not hold the same ranks. Men tend to hold higher ranks than women and to work at more prestigious research universities. Whatever their productivity, women's achievements are not equally rewarded by salary increases, promotion, or professional recognition.[40]

This finding alone goes a long way toward solving the so-called productivity puzzle. Women holding lower ranks at less prestigious universities command fewer resources. Men as a group outproduce women statistically at such a high rate because a few well-placed men turn out large numbers of papers. These men benefit from what sociologists call "cumulative advantage"—those who do well professionally amass the resources to do even better in the future. Men are more likely to be among the academic elite, those with endowed chairs, generous funding, spacious and modern labs, collaborators around the world, membership in national and foreign academies, and prestigious prizes. Women, by contrast, tend to have more difficulty getting connected to that world; they suffer from "cumulative disadvantage," or subtle, unquantifiable discrimination.[41]

Citation Counts

Research performance is measured in two ways: the number of papers published and the number of times a particular paper is cited. The sheer number of publications tells nothing about the quality or impact of a scientist's work. Many papers are published to satisfy granting agencies or for promotion or salary increases and may contribute little to human knowledge. Over half of all papers published are never cited, and 80 percent are cited no more than once (and sometimes these single citations are an author citing his or her own work). Consequently, citation counts (weighted by the prestige of the journal—10 percent of science journals are cited 90 percent of the time) are used to evaluate the importance of a person's work.[42]

Men publish more papers, hence their overall citation rates are higher. Paper for paper, however, women's are cited at almost the same rate as men's (5.02 citations on average for a woman in 1984, 4.92 citations on average for a man). More recently J. Scott Long astonished many with his finding that, among biochemists, the average paper by a woman was cited 1.5 times more often than the average paper by a man.[43]

Why, despite women's more marginal location in the academy, should their papers have as much, and perhaps even more, impact than men's? Gerhard Sonnert and Gerald Holton have found that women have different standards for publication than men: the women they interviewed said they value thorough and comprehensive research; fewer men characterize their publications this way.[44] How do we account for this discrepancy? Are women less aware of the need to publish quickly and often? Are they "less strategic" in their approach to publishing, as Sonnert and Holton suggest?

If there is a tendency for women to produce more significant work this, oddly enough, may be a legacy of discrimination. Because their presence in science is often challenged, women may hesitate when it comes to publishing. According to Susan Gerbi, a cell biologist at Brown University, "women tend to be a bit more insecure about how their work will be received . . . and want an airtight story, a more complete story, before they go public." Being more thorough slows productivity. Long offers a different explanation: not that women are more exacting than men concerning publishing but that they have different positions in the scientific community. Men are more often senior scientists who, in addition to their own work, sign their names to a number of less significant papers in their role as dissertation director or lab director. While this inflates their productivity, it deflates their citation rates. Sonnert and Holton found that women, by contrast, often prefer to carve out a niche of their own and work on their own problems rather than join the heated competition on breakthrough topics.[45]

Counting publications and citations does not take into account many forms of structural discrimination. Several classic studies of "what's in a name" have revealed that academic culture values men's work over women's. Studying readers' responses to names, psychologists have shown that, even when the content is the same, readers prefer articles said to be by men. Researchers gave both men and women articles by authors identified variously as John T. McKay, Joan T. McKay, J. T. McKay (supposedly sex-neutral), Chris T. McKay (ambiguous in regard to sex), and Anonymous. The articles were identical in all ways except for the name of the supposed author. Both men and women rated an article more favorably when it was attributed to a man rather than a woman. Moreover, they preferred the ambiguous "J. T." over "Joan," but not over "John." Readers rated an article significantly lower when they thought "J. T." was a woman attempting to hide her identity. In general, readers treated "J. T." more like a woman than like a man.[46]

Volumes could be written on naming practices. Women in the United States who use the names given them at birth are often mistakenly called "Mrs." This intended show of respect conjures instead an image of incest that the speaker surely does not mean (to call me Mrs. Schiebinger implies I am my father's or my brother's wife). The term "Ms.," designed as a parallel to "Mr." and not requiring knowledge of marital status, never caught on in the society as a whole because it was too closely identified with feminism.

The readers in the study of reactions to authors' names were perhaps correct to assume that J. T. was a woman seeking to hide her identity: women sometimes do try to protect themselves by using initials, as in the phone book. Naming practices also differ by discipline. In the physical sciences, where women are not well represented, authors are most often identified by initials alone (perhaps because of the large number of coauthors). In the 1960s, when women were a rarity in these fields, some physical science journals exempted women from this practice, either allowing them to use their first names or identifying them clearly as women.[47] In the humanities, where women are more abundant, authors commonly use a formal first name in addition to a middle initial.

Despite their precision, publication and citation counts do not tell us much about discrimination in the sciences. Scholars in this area tend to focus on how women might become more competitive by making choices concerning marriage, job mobility, and patterns of collaboration that lead to greater success. They do not take into consideration the many subtle barriers that still tend to disadvantage women. As we shall see in Chapter 4, women still experience discomfort in the culture of science. Moreover, the *Science Citation Index* rarely includes sources in languages other than English, so the hair-splitting measures of productivity and influence do little to consider science in a global context.[48]

Surveys

Statistics and publication and citation counts are all attempts to measure equity. Surveys are attempts to quantify incidents of discrimination. The amount of dissatisfaction expressed in surveys of women scientists is surprising. It may be that women who might not complain publicly for fear of rocking the boat feel freer to do so anonymously. The terms "discrimination" and "harassment" also mean different things to different people. It is difficult to determine to what extent overreporting is a problem. I, like many others, tend to believe that overt discrimination or harassment (girlie calendars in public places or sexual innuendo) is a thing of the past in the professional world. But many women report otherwise. In 1991 the American Astronomical Society found that 40 percent of its women members felt that they had experienced or witnessed discrimination, while only 12.4 percent of the men surveyed said they had ever witnessed any form of discrimination against women astronomers. In a second survey 39 percent of the society's women members reported having been taken less seriously than their male colleagues. Similar problems were revealed in a 1992 survey of the American Chemical Society: 43 percent of the women reported having encountered sexual discrimination in the workplace. In 1993 the *New England Journal of Medicine* indicated that nearly three-quarters of women students and residents are harassed at least once during their medical training, and that three-quarters of the women doctors surveyed had been sexually harassed by male patients. Even the women science and engineering students polled in 1991 said they dealt daily with "the irritation of open (or thinly veiled) sexist remarks from their male peers, and with the inner stresses of feeling unwelcome and pressured."[49]

Many successful women feel shut out of the real centers of power. "It is easy for women to share power with men on committees as long as those committees are not powerful," says Patricia Goldman-Rakic, a professor of neuroanatomy and neurophysiology at Yale University. "But when it is near the center of power . . . the committee is likely to be all male." Barriers are maintained in part because even well-meaning men and women tend to know more persons of their own sex and to think of them first when putting together committees, conferences, or other working groups. Linda Maxon, a former head of Pennsylvania State University's biology department, has remarked, "It's hard to call it prejudice; it's just human nature." Conscious efforts to include women (or in some instances men) can improve the situation. In 1988, when the Princeton professor Shirley Tilghman organized a Gordon Conference on molecular genetics, about 33 percent of the speakers and 45 percent of the participants were women. Two years later, when a conference on the same topic was arranged by a committee consisting only of men, only 2 speakers out of about 100 were women.[50]

More common than outright harassment is the steady diet of small offenses and innuendo that some women endure. Dr. Frances Conley, a top neurosurgeon, who resigned from Stanford University's Medical Center after twenty-four years of service, told the press in 1991: "I resigned my position as a full, tenured professor because I was tired of being treated as less than an equal person. I was tired of being condescendingly called 'Hon' by my peers, of having my honest differences of opinion put down as a manifestation of premenstrual syndrome, of having my ideas treated less seriously than those of the men with whom I worked . . . I resigned because of a subtle sexism that, while not physically harmful, is extremely pervasive and debilitating." She described an environment where, as late as 1991, faculty members spiced up lectures with slides of *Playboy* centerfolds, where sexist comments were frequent, where those who were offended were told to be less sensitive, and where unsolicited touching and fondling occurred between male house staff and female students. At age fifty, Conley said, she no longer wished to work in a "hostile" environment. She subsequently rejoined the faculty after the man she considered one of the worst offenders stepped down as chair of her department and agreed to undergo sensitivity training. Despite the increasing number of women in medical school, medicine continues to hothouse sexism within its rigidly structured educational hierarchy.[51]

Legislation concerning sexual harassment has been put in place primarily to protect women. But calling attention to harassment can deepen divides between the sexes. As Beverly Sauer put it, "Like toxic chemicals, sexual harassment can poison a workplace, creating an atmosphere of distrust and suspicion."[52] In the early 1980s, just as sexual harassment was becoming a public issue, a Harvard University professor stopped having lunch with his unmarried women doctoral students in order to avoid possible misunderstandings. While his precaution supposedly kept the unmarried women out of harm's way (there is no evidence that this particular professor had ever posed a threat), it also excluded them from informal contact with their adviser.

Efforts to avoid the appearance of indelicate behavior can interfere with the building of the strong working relationships crucial to professional success and congenial working conditions. A woman graduate student noted a similar distance between men professors and women students at

Stanford, "I don't go hiking with my adviser. I don't lunch with my [male] adviser. I don't bullshit with my adviser like the boys do." Invisible barriers surrounding sexuality continue to divide people, male or female, heterosexual or gay. In order to allay what she perceives as anxieties about her sexuality, a lesbian professor of color said, she makes it a point never to go anywhere—to coffee or lunch—with an individual student, straight or gay. When a meeting is absolutely required, she takes the student to a café directly across from her office and makes sure that the student takes out papers and reading materials as a visible sign that their meeting is professional in nature.[53]

One still finds examples of blatant sexism today, but less often than in the past. More interesting are the biases against women—many unintentional—that persist among even well-meaning people. Men and women working in the same institution, teaching or studying in the same department, often have very different experiences. Women still encounter a host of subtle personal and social barriers—barriers that productivity counts do not uncover and that laws alone cannot remove. These barriers are often so much a part of the everyday way men and women relate to each other that they may not even be noticed.[54] Then, too, women can perpetuate their own subordination when they engage in stereotypically feminine behaviors (though this is often expected and difficult to avoid). As we shall see in subsequent chapters, attitudes toward gender are not peripheral to science but structure key aspects of both the institutions in which science is produced and the knowledge issuing from those institutions.

personal and social barrier

*Ⓒ 3

The Pipeline

IN THE 1970s government officials and university scholars tended to approach the problem of women's low participation in science from the top down: various discriminatory practices were seen as blocking women's paths as they attempted to rise through the ranks. In the late 1980s the problem was reconfigured in terms of a "pipeline," with all the unattractive connotations that metaphor suggests. The new model looked at things from the bottom up, predicting that if more girls entered the educational end of the pipeline, more women would be turned into credentialed specialists and empty into the science job pool. The problem was viewed less as one of discrimination than as one of self-(de)selection: too many girls were opting out of math and science at too early an age. The analysis assumed a solution. Women or, better, girls needed to be given more training and encouragement in science. Pipeline liberals saw the solution to the low numbers of women in science as one of reforming the individuals—giving girls the benefits of boys' socialization.

The statistician Betty Vetter painted an arresting, if somewhat teleological, picture of the pipeline. Of any 2,000 ninth-grade boys and 2,000 ninth-grade girls, only 1,000 of each group have sufficient training in mathematics to continue in science. At the end of high school, only 280 of the men and 210 of the women will have taken enough math to pursue a technical career. In college, 143 men but only 45 women will major in science. Once women have chosen to major in science, a larger percentage of them complete their degree than men: 44 of the original men and 20 of the original women will graduate. Women go on to graduate school in the same relative proportion as men, but many stop at the master's level.

Slowed to a trickle, from the original 2,000 students in each group, the pipeline will yield 5 men but only 1 woman with Ph.D.'s in some field of the natural sciences or engineering.[1] In other words, it takes 400 ninth-grade boys to get one Ph.D. scientist but 2,000 ninth-grade girls.

Scholars looking at the scientific pipeline generally assume that a child's environment is a major factor in refining skills and fostering future interests. Factors that will lead girls to reject science as a career are thought to be cultivated very early—even moments after birth. In one study, parents were asked to describe their newborn babies—at a time when one of the few things they knew about the child was its sex. Parents proudly described many of the boy babies as active and exploratory, while they delighted in thinking of the girls as small, soft, and delicate. Another study found that adults tend to give children toys that reinforce sexual stereotypes. When an infant was introduced to a group of adults as a girl, they tended to give her "feminine" toys, such as dolls and stuffed animals, and talked to her more. When the same infant was introduced as a boy, many adults offered "masculine" toys, such as balls and cars, and played more rough-and-tumble games. Given that adults tend to choose toys along gendered lines long before the children themselves can express a preference, it is not surprising that by eighteen months of age, when children begin to pick their own toys, they often choose those which are familiar. Giving girls and boys different toys might be harmless except for the fact that toys create aspirations, hone conceptual skills, and encourage certain behaviors to the exclusion of others.[2]

Many enlightened parents and playschool teachers attempt to give dolls to boys and building-blocks to girls. My partner and I discussed raising our first child "gender-free"—whatever that might mean—when he was born in 1989, but such things are difficult if you live in a society. Cultural pressures can overwhelm the best intentions. Toy manufacturers, for one, play heavily on gender stereotypes in their designs. An ad published in 1969 in *Life Magazine* announced: "Because girls dream about being ballerinas, Mattel makes Dancerina . . . a pink confection in a silken blouse and ruffled tutu . . . Because boys are curious about things big and small, Mattel makes SuperEyes, a telescope that boys can have in one ingenious set of optically engineered lenses and scopes." Toys are no less gendered today. Toy catalogs picture boys with building sets, Nerf guns, alien monsters, and micromachines, and girls with Barbie and her many accessories, stuffed animals, and toy makeup kits. Manufacturers insist that toys be clearly gendered. In 1989, when Jaron Lanier was developing the data glove for Nintendo games, he strongly resisted sex typing. Toy manufacturers, however, immediately cast the glove as a boys' toy and embellished it with black Darth-Vaderish, sports-car-like paraphernalia. Had it been designated for girls, it would no doubt have been pink and frilly. Toy manufacturers have recently released a new set of pink and pastel Legos in an attempt to capture the female market.[3]

As anyone who has visited preschool classrooms (where typically the girls play dress-up in one corner while the boys build with Legos in another) can tell you, children begin forming their own culturally sanctioned sexual stereotypes as early as age two. Girls often say they want to become nurses or teachers, while boys exude enthusiasm for becoming police officers, sports superstars, garbage men, or doctors. In a culture that gives preference to things masculine, girls today may say they want to become "police girls" (presumably the female equivalent of "policemen"), pilots, or lawyers. But boys rarely choose from the traditionally feminine side of life, seldom stating a burning desire to become a nurse, a househusband, or an elementary school teacher. Children's books continue to reinforce sexual stereotypes. A hefty three-quarters of recent prizewinning children's books portrayed women doing housework and men working outside the home.[4]

Gendering continues in grade school. The mission of U.S. public education has always been to promote equality through equal opportunity. But boys and girls receive very different educations even when sitting in the same classrooms and studying the same curriculum. From preschool to university, teachers tend to choose classroom activities that appeal more to boys than to girls. Sociologists videotaping math classes have found that, usually without realizing it, teachers often give boys more freedom to discover alternative solutions to problems, while encouraging girls to follow rules more closely. They may demonstrate a problem for the girls, while expecting boys to figure it out for themselves.[5] One study of fourth- and fifth-graders revealed that boys were praised for intellectual capabilities, while girls were more often praised for neatness. Boys tend to demand more attention of teachers, calling out and guessing answers. In classes where teachers talk more to boys, girls become quieter as time goes on, closing up, as one author suggests, "into a shell." Girls are generally quietest in classes where they are most in the minority. The net result is that U.S. math teachers—both men and women—teach girls less math than they teach boys.[6]

Through all of this, girls as a group earn higher grades than boys. Some say girls receive better grades because they hand in all assignments and complete their work carefully, suggesting that girls excel because they are good citizens rather than because they are talented. Interviews indicate that teachers think of girls as conscientious, serious, quiet, and self-motivating—surely traits worthy of reward and often predictive of future success.[7]

The same tendency to privilege boys over girls can be seen in computer software. Researchers have made the disturbing discovery that much of the educational software, designed to teach the fundamentals of math, spelling, and language, appeals more to boys than to girls. It is not the computer or the software but the expectations of the programmers (many of them women) that have created these gender asymmetries. Software designers tend to assume that users will be male. Thus educational software, written for the generic "student," is in fact designed for boys. Take, for example, *Demolition Division,* a popular game described by its producer as "an opportunity to practice division problems in a wargame format." Students fire guns at correct answers (posted on tanks) that move across the computer screen. Hits and misses are recorded at the bottom of the screen. Boys especially like action-oriented games with loud noises, flashing colors, and hand-eye coordination that require quick reflexes and aggressive reactions.

Girls often find "blasting asteroids out of the sky" boring and prefer word-oriented software, practical tasks, puzzles, and mazes. When confronted with software designed for boys, girls experience stress that can lower their motivation and performance. Girls report feeling the most stress when working with cross-gendered software in public settings. Boys using software designed for girls also feel anxious, but they are less often confronted with the problem. Companies have responded by producing software just for girls: *Barbie Super Model, Beauty and the Beast: Belle's Quest,* and *Let's Talk about Me!* Some of the newest software for girls features a multicultural group of female characters who can be guided through a complex set of personal encounters. No princess is rescued or glittering treasure recovered; rather a better sense of self supposedly emerges from secret treehouse discussions of family, friends, and feelings. While none of these games requires anyone to be shot, disemboweled, or blown up, they do have their own gender hazards.[8]

Girls, then, are not getting the same education as boys in American schools, particularly in mathematics, which is considered the "critical filter" determining whether or not women go into science and engineering careers.[9] Boys and girls display similar mathematical interests and abilities until about seventh or eighth grade, when many girls begin losing confidence in their mathematical skills and elect fewer math classes.

Scholars have begun to correlate this drop in mathematical confidence with a general drop in girls' self-esteem. In one study 70 percent of elementary school boys and 60 percent of girls responded positively to the question "Are you happy with the way you are?" In high school, half of the boys continued to be satisfied with themselves and their achievements, while 70 percent of the girls expressed grave dissatisfaction with some aspect of their appearance, personality, or ability. This lack of confidence was especially marked among girls studying math and science. Another drop in self-esteem occurred in the transition from high school to college. In a study of high school valedictorians (46 women and 34 men), the men and the women expressed about the same degree of self-esteem in their senior year of high school. By the end of their senior year of college, however, not one of the women rated herself as having intelligence "far above average," while one-quarter of the men did—despite the fact that the women's grade point averages were higher overall than the men's.[10]

It is a sad fact of American life that women often underestimate and men overestimate their abilities and probability of success. I was in graduate school before I learned that men tend to exaggerate. I learned that they exaggerated everything: their height, their success, their prospects. I also learned that I had to put my own achievements in the best light when writing a curriculum vitae and letters of application. Being a professional these days, in the United States more than elsewhere, requires what an MIT graduate student has described as "strutting behavior." This is as true of science as any other field.[11]

But society expects women, more than men, to be modest, and many internalize this imperative early in life. This is especially alarming because low self-esteem is a correlate of modesty. In a study of undergraduate students three-quarters of the women, compared with less than half of the men, cited low self-esteem as their reason for leaving science. While young women's self-estimations may not always be consistent with their academic performance, they are consistent with other of their experiences. Scholastic Aptitude Test (SAT) scores, for example, designed to predict success at university, underpredict women's grades and overpredict men's (see also Chapter 9).[12]

Even women who have distinguished themselves in science sometimes suffer from a form of self-doubt that Sheila Widnall, a past president of the American Association for the Advancement of Science and former Secretary of the Air Force, has called the "impostor syndrome." Mildred Dresselhaus, a member of the National Academy of Sciences, has said, "When I started out, I felt all the time like an amateur." She was surprised to be offered a position as a full professor at MIT, and remembers thinking "Why me?" She had been doing science just for the fun of it and had not thought of herself as a professional. In 1946, when the Nobel Prize winner Maria Goeppert Mayer was offered a part-time job at Argonne National Laboratory, she responded, "I don't know anything about nuclear physics." A 1982 survey of 500 British women scientists revealed that many harbored similar feelings of inadequacy and self-doubt. A 1995 study of high-achieving women scientists showed that only half saw their own scientific ability as being above average (compared with 70 percent of the men).[13]

The literature on why women leave science has emphasized different cultural expectations for boys and girls, gender inequalities in education, and the potentially devastating effects of the isolation experienced by women in careers traditionally reserved for men. The sociologists Stephen Cole and Robert Fiorentine, however, argue that women's poor representation in science results not from unequal opportunities but from a "persistence gap" between men and women. Cole and Fiorentine argue that it is now acceptable for women to pursue high-status careers, like medicine, business, or law, but that women fail because they do not try hard enough. Women, they claim, are less persistent in their careers than men because they can rely on the socially sanctioned safety net of marriage. Men, by contrast, achieve social standing almost exclusively through occupational success. Thus men must persist in high-prestige occupations even in face of adversity, while women need not try as hard because if they fail professionally they can always become somebody's wife. Gerhard Sonnert and Gerald Holton's study of elite scientists bore this out to some extent: 80 percent of the men but only 34 percent of women said they were the primary wage earner in their family.[14]

Who, then, are the women who stay in science? Surveys of women choosing science majors at Harvard and Stanford show that they come from wealthier and more highly educated families than the men studying science at those institutions. They also typically come from families with fathers in scientific or technical occupations. Perhaps the most important fact about women who stay in science is that they are very talented. Only women with exceptionally high SAT scores and GPAs major in science.

Because women come under close scrutiny, they develop extremely high standards for themselves as a prerequisite for going into and staying in science, sometimes feeling that they must outshine the men.[15]

Women who succeed in science also tend to be graduates of single-sex schools; this is true in the United States but also in many European countries. In the United States, liberal arts colleges, and particularly women's colleges, produce a disproportionate number of women who continue working toward Ph.D.'s in math, science, and engineering. What is the secret of women's colleges' success? For one thing, in the absence of men women students exercise greater leadership, routinely taking charge in laboratory exercises and classroom discussion. Further, almost half of the math and science faculty at U.S. women's colleges are women (45 percent compared with 11 percent at coed institutions and 5 percent at technical institutes). Women students thus have ready role models and mentors. Similar patterns hold for African Americans—both men and women. Historically black colleges and universities (founded as schools for freed slaves) awarded 40 percent of all African-American B.S. degrees in the natural sciences in 1989. At Spelman College, traditionally an African-American women's college, 37 percent of the students major in math or science, and half of the women go on to graduate study. Traditionally black colleges and universities are the primary source of African-American students who pursue doctoral degrees in science. Students say these colleges provide an atmosphere in which race is not an everyday issue and build confidence among students.[16]

U.S. women's colleges, founded in the late nineteenth century when East Coast all-male universities refused to admit women (the "Seven Sisters" are sisters to the Ivy League schools), are celebrated today for having all the elements in the currently approved recipe for "female-friendly science." They provide a critical mass of women students and professors, thus overcoming the isolation that commonly afflicts women in traditionally male fields. Science courses at women's colleges are rarely designed to weed out or intimidate; "female-friendly" classrooms play down competition and foster cooperative learning. Teachers, cast as guides, mentors, and question-askers rather than as knowledge-givers, discuss the practical application of scientific problems, their social origins and consequences. The goal, as characterized in the literature, is to help students see the relationship between what is known and what is questioned. At Spelman College, as Etta Falconer, the director of science programs, remarked, "We expect our students to succeed and they do."[17]

Other factors may also play a role in the success of women's colleges in producing scientists. First and foremost, many are elite institutions, where students have not only the advantage of abundant resources but also the confidence that accompanies a lifetime of excellent education. Second, these colleges expect faculty members to devote more time to teaching than do universities, where the emphasis is more often on research. In such a setting, female and male students alike receive more individual attention.

Women's colleges (and historically black colleges) also have some distinct disadvantages, especially for faculty members. Not being primarily research institutions, they rarely provide the kinds of equipment and laboratory support crucial for advancing faculty members' careers.

The success of women's colleges has sparked discussions about replicating their design in other contexts. In the United States several coeducational universities have made various all-female settings available to women students. While it is currently illegal for publicly financed schools and universities to offer sex-segregated courses, differently configured sections of introductory and upper-level science courses—some exclusively for men, some exclusively for women, and some mixed—provide a variety of opportunities for students and their varying educational needs.[18]

Critics argue that the single-sex approach does not confront prejudice head on and only delays girls from learning to maneuver in male-dominated environments. Women scientists who attended all-girls schools differ in their responses to the experience. Some found it helpful not to have to compete with men until graduate school, when they already felt confident about their abilities. Others found single-sex schools oppressive and unhealthy in that they tended to exaggerate the strangeness of the opposite sex, turning men into exotic creatures met only on weekends.[19]

Even after women have made it through graduate school and landed a job, they continue to "leak" from the pipeline. They are twice as likely as men to leave careers in science and engineering. Between 1982 and 1989 more than 20 percent of all women working in science and engineering left their jobs. The situation in industry is worse: women in industry walked out on their jobs twice as often as women in the public and nonprofit sectors. They leave for a variety of reasons: not being invited to professional meetings, having their performance judged by different standards than men's, having to work harder to have their work valued as highly as a man's. These women mention the struggle to balance family and career, the need to hide pregnancies as long as possible, inflexible working conditions, and an environment in which some employees compete to see who can put in the longest days. They note the difficulties women face advancing into management, salary discrepancies, and the disrespect embedded in allegations of reverse discrimination.[20]

I have focused on cultural forces that discourage girls and women from pursuing science. Naturists would interpret things quite differently, emphasizing that behavior or career choice also reflects natural differences between the sexes. Studies of CAH girls (who were exposed to abnormally high androgens in the womb), for example, have shown that when given a choice of typically masculine toys (trucks, cars, guns) and typically feminine toys (dolls, kitchen supplies, board games), the CAH girls prefer the more masculine toys. For centuries naturists have attributed intellectual differences between the sexes primarily to natural causes, whether those be the heat and dryness of the body (Aristotle and Galen), the size of the cranium (Le Bon), natural and sexual selection (Darwin), hormones (Edward Clarke), or brain asymmetries (Kimura). These differences are generally taken to imply that men and women will have different professional interests, and that women, who have weaker spatial and math abilities, will not be equally represented in engineering or physics.[21] Despite extensive research, it has not been determined to what extent sexual differences are due to environment or to genetic makeup. The best we can do is to remove any lingering cultural impediments to women's success in science.

Most scholars focus on why women leave science, suggesting that those who leave have somehow failed. They have failed to take enough math courses, to persist under duress, to publish enough. These scholars rarely look at successful women and why they, too, sometimes leave science. The pipeline model does not account for successful women who, after achieving professional positions, choose to leave.

Some successful women, like Evelyn Fox Keller, leave laboratory science to study the history and philosophy of science. Others, like the astronomer France Cordova, leave active research to make a mark in national policy on science. Sadly, concern for the historical origins of priorities in science or for public policy rarely fit comfortably into full-time laboratory science. Other successful women leave because they become discontented with the social implications of their research.

Martha Crouch of Indiana University has explained her reasons for leaving molecular biology in 1990. She, like many others, had been drawn to science by the crucial importance of basic research. For many years she worked to understand the role of maternal contributions to the maturation of plant embryos. After some time she realized that her work was not "pure" research but also was useful to certain agribusinesses, such as the palm oil industry. While the questions guiding her research were interesting in their own right, Crouch judged that the reasons they, and not other equally interesting questions, were asked had to do with the need to produce plants with reliable high yield. While her project contributed to the production of vegetable oil and increased agricultural income for peoples of tropical countries, it also had negative effects—such as the displacement of small farmers and the degradation of the environment—that Crouch found unacceptable. Because she could not continue her research without furthering these negative consequences, she left laboratory science to study rural economics, history, political science, and ecology in an effort to implement what she considered a more socially responsible plant science.[22]

Regine Kollek is another who left. Drawn to molecular biology, Kollek dreamed of understanding the secrets of life and, more practically, hoped to help cure hereditary diseases. She joined a group at the University of Hamburg, in Germany, to investigate how certain viruses might be used to fight cancer. To this end her lab began creating new viruses that can cross species boundaries. Such viruses have the potential to infect humans—for Kollek, an unacceptable risk. She discussed her concerns with the group leader, whose response was that if the group did not do the research someone else would. After a year their dispute was resolved at a higher level and certain safeguards were introduced into the lab. Still, Kollek believed the work was too dangerous to human populations and left the lab in 1984 to turn her attention to the ethical and political problems surrounding gene manipulation.[23] Her concerns were branded political and seen as lying outside the normal course of science. Today Kollek directs the division for biotechnology, society, and environment at the University of Hamburg, a newly created division to consider the long-term consequences of biotechnology.

For the past decade the pipeline model has informed many government, university, and industry policies and led to numerous intervention programs aimed at keeping more women on track in science. While interventions are essential stopgaps, they alone cannot solve the fundamental problems distancing women from careers in science. Intervention programs address problems piecemeal—attempting to provide mentors in an atmosphere of isolation, introduce maternity leave to institutions modeled on men's life cycles, hold girls' interest in math in classrooms designed around boys, reform hiring and promotion practices through affirmative action—and as such cannot change deep and structural patterns of discrimination. A 1994 National Research Council report showed that the model is flawed: in medicine and business, where women have been in the pipeline for twenty to twenty-five years (long enough to emerge into top positions), the model has been proven incorrect. The pipeline model, built on the liberal assumption that women (and minorities) should assimilate to the current practices of science, does not provide insight into how the structure of institutions or the current practices of science need to change before women can comfortably join the ranks of scientists.[24]

II

GENDER IN THE CULTURES OF SCIENCE

Math class is tough.   BARBIE, 1992

The requirements of correctness in practical judgments and objectivity in theoretical knowledge . . . belong as it were in their form and their claims to humanity in general, but in their actual historical configuration they are masculine throughout. Supposing that we describe these things, viewed as absolute ideas, by the single word "objective," we find that in the history of our race the equation objective = masculine is a valid one.

GEORG SIMMEL, sociologist, 1911

One of the most serious problems women . . . have is conceptualizing and acting upon the subtle non-articulated lack of acceptance.

KAREN UHLENBECK, mathematician, 1997

ᶜᶜ 4

The Clash of Cultures

W HEN Barbie, America's emblem of hyperfemininity, uttered her first words in the summer of 1992, she told the most recent of her 800 million owners that "math class is tough." After protest by women's groups, the makers of Barbie (the CEO of Mattel at that time was a woman) removed this statement from the doll's repertoire of ready phrases.[1] But why did they assume that their platinum bombshell with the hourglass figure and the feet permanently imprinted for impossibly high heels would find math difficult?

Does science have a gender? Many have argued that it should. Sir Francis Bacon, the seventeenth-century English ideologue, called for the Royal Society of London to "raise a masculine philosophy" (as the new science was called). The nineteenth-century German historian of philosophy Karl Joël, appalled by what he saw as the excesses of the French Enlightenment, urged a return to manly *(männliche)* philosophy and applauded the arrival of a masculine epoch *(Manneszeitalter)* ushered in by the critical philosophy of Immanuel Kant. (Kant taught, among other things, that anyone engaged in serious intellectual endeavor should have a beard.) Even the great English feminist Mary Wollstonecraft, in her efforts to create equality between the sexes, encouraged women to become "more masculine and respectable."[2]

In our own century, Georg Simmel argued that objectivity, though seemingly applying to humanity, was in fact an attribute of masculinity. In 1985 Evelyn Fox Keller, rephrasing Simmel, declared that science is "masculine," not only in the person of its practitioners but in its ethos and substance. In the spring of 1993 the magazine *Science* asked, "Is there a 'female style' in science?" This was a rephrasing in essentialist language of a question posed in the 1980s: "Is there a feminist science?" Both questions imply that scientific culture exhibits the markings of masculinity.[3]

Today, bald declarations that science is masculine raise ire among many scientists. The discussions about gender in scientific cultures that began in the 1980s, however, turn attention away from women—their triumphs, trials, and tribulations—and the notion that they simply needed to perform better in the world of science. These discussions have instead brought critical attention to the cultures of science and how gender continues to distance women from the professional world of science.

In Chapter 1 we looked at the history of women's engagement in scientific institutions—universities, scientific academies, and so forth. A culture is more than institutions, legal regulations governing a profession, and a series of degrees or certifications. It consists in the unspoken assumptions and values of its members. Despite claims to value-neutrality, the sciences have identifiable cultures whose customs and folkways have developed over time. Many of these customs took form in the absence of women and, as we shall see, also in opposition to their participation. How have the cultures of science, with their rituals of day-to-day conformity, codes governing language, styles of interactions, modes of dress, hierarchies of values and practices, been formed by the predominantly male practitioners of science? What, in other words, is the historical relationship between gender and science?[4]

My purpose here is not to judge the virtues of femininity or science but to highlight the historical clash of these two cultures. While it is abundantly clear that women—of various classes and ethnic backgrounds—do not share one culture, science also has many cultures and subcultures. Nonetheless, women who become scientists in the United States or Europe often live in two worlds—the world of science and the world of womanhood—with very different expectations and outcomes. Strategies for success learned in one world can be lethal in the other.

There are certain dangers in calling attention to gender differences. The law professor Martha Minow identified in 1984 what she called "the difference dilemma"—that calling attention to gender stereotypes can reinforce them and create friction where before there seemed to be none, but that ignoring gender differences can leave invisible power hierarchies in place.[5] We also know that "masculinity" and "femininity" do not have universal meanings above and beyond historical contexts. These terms mean very different things at different times and in different places, and they often refer as much to the manners of a particular class or a particular people as to the characteristics of a particular sex. For the founders of the Royal Society of London in the seventeenth century, the much-trumpeted new and "masculine" philosophy was to be distinctively English (not French), empirical (not speculative), and practical (not rhetorical). "Masculinity" served in this case as a term of approbation, having nothing to do with women and attaching only tangentially to men. "Masculinity" and "femininity" do not map directly onto sex (nor should they). As decades of scholarship have demonstrated, however, sexual differences define powerful fault lines in our culture. Many of the differences between men and women I will discuss are historically real. Many gendered behaviors come to us so naturally (we learned them long ago and well) that we engage in them unconsciously. That does not mean that they are necessarily desirable or that every man or woman fits a stereotype.

Gender in the style of science is significant because women's long legal exclusion from scientific institutions was buttressed by an elaborate coding of behaviors and activities as appropriately masculine or feminine. Unearthing assumptions surrounding gender in science helps unearth unspoken notions about who is a scientist and what science is all about and how these notions have historically clashed with expectations about women. Understanding gender in the professional world of science can help cultivate new behaviors and solidify good relations between the sexes within universities, industries, government, and domestic life.

The Gendering of Science

The ardent gendering of science developed at the end of the eighteenth century as women were being forced from the newly formalizing scientific institutions. Women did not go quietly. Elaborate cultural prescriptions for science and for women accompanied women's formal exclusion from science, making this exclusion seem natural and just. Only within this context can we understand the urgency with which Europeans and Americans cultivated opposing ideals of science and of femininity.

Two key developments in European science and society—the privatization of the family and the professionalization of science—were crucial in structuring this historic clash of cultures. The Enlightenment was a time when European society was being rebuilt: all men, the slogan taught, are by nature equal. But not all men and certainly very few women were to become equal participants in what came to be defined as the public

sphere of life. In the seventeenth and increasingly in the eighteenth century, European society diverged politically and economically into two
separate spheres: the public sphere of government and the professions and
the private sphere of hearth and home. Men (elite and middle-class men)
found their "natural" place in the public sphere while women of those
classes became newly empowered mothers within the home.

If the new rights of citizens were not to extend to women, liberal democratic theory had to be amended. The theory of *sexual complementarity*—that women are not the equals of men but their complementary op-
posites—fit neatly into dominant strands of liberal democratic thought,
making inequalities seem natural while satisfying the need of European
society for a continued sexual division of labor. Henceforth women were
not to be viewed merely as inferior to men but as fundamentally different
from, and thus incomparable to, men—physically, intellectually, and mor-
ally. The private, caring woman emerged as a foil to the public, rational
man. As such, women were thought to have their own part to play in the
new democracies—as mothers and nurturers.[6] Complementarians sought
to eliminate competition between men and women in the public sphere
by removing women from that sphere.

This new doctrine carried with it the answer to the question of women's
participation in science. For complementarians, the purposes and activities of the public realm differed essentially from those of the home. As
the great German philosopher Georg Wilhelm Hegel put it, in the state
everything originates in abstraction, in concepts; while in the home everything originates in the physical needs of heart and soul. Family piety,
or the law of inner life, Hegel continued, was the law of woman. This
law, based in subjectivity and feeling, stood opposed to the universal character of the public law of the state.[7] Science was part of the territory that
fell to the masculine party in this restructuring of eighteenth-century culture. Because science, like any other profession, came to inhabit the public
realm where women (or femininity) dared not tread, science came to be
seen as decidedly masculine.

Complementarity developed with the enthusiastic participation of the
scientific community (see Chapter 6). Within this framework, femininity
came to represent a set of qualities antithetical to the ethos of science.
The ideal virtues of femininity—required for the joys of domestic life—
were portrayed as personal failings of women in the world of science. A
growing number of anatomists and men of science held that creative work
in the sciences lay beyond the natural capacities of women: women, mired

as they were in the immediate and practical, were incapable of discerning the abstract and universal. Women lacked genius: they could succeed in small works which required only quick wit, taste, or grace; they could even acquire erudition, talents, or anything else which was acquired as a result of work. But their work was only cold and pretty, for women lacked genius—that "celestial flame" which warms and sets fire to the soul. Participation in science required a certain strength of mind and body that women simply lacked. In the nineteenth century Francis Galton pronounced men of science "strongly anti-feminine; their mind is directed to facts and abstract theories, and not to persons or human interests . . . they have little sympathy with female ways of thought."[8]

In defining why women could not do science, complementarians were not defining women so much as what was unscientific. Women—as representatives of private life—were repositories for all that was not scientific: in a scientific age women were to be religious; in a secular age they were to be the keepers of morals; in a contractual society they were to provide the bonds of love. Complementarians conceived femininity as a necessary counterbalance to masculinity: each gender was incomplete in itself, but together they constituted a workable whole.

The Enlightenment idealization of women as the angels of the home applied, however, only to middle-class Europeans. Neither the dominant theory of race nor that of sex in this period applied to women of non-European descent, particularly those of African descent. In 1815 Georges Cuvier, France's premier comparative anatomist, performed his now infamous dissection of a South African woman known as Sarah Bartmann. The very name Cuvier gave this woman—*Vénus Hottentotte*—emphasized her sexuality. In his memoir he made it very clear that Africans were not included among those who could do science: "No race of Negro produced that celebrated people who gave birth to the civilization of ancient Egypt, and from whom we may say that the whole world has inherited the principles of its laws, sciences, and perhaps also religion."[9] Like other women, Sarah Bartmann did not fit comfortably in nineteenth-century racial hierarchies where primarily men were studied for their comparative superiority. Like other Africans, she did not fit European gender ideals. Elite European naturalists who set such store by sexual complementarity when describing their own mothers, wives, and sisters did not include African women in their new definitions of femininity.

I should note again that there is nothing natural or necessary about these characteristics defined by Western cultures as feminine or as scientific. Ideals of masculinity, femininity, and science developed historically, informed by and responding to the economic need to have women serve as household managers and men work outside the home, and to the political desire to have only property-owning men vote in participatory democracies. Gendered characteristics—typically masculine or feminine behaviors, interests, or values—are not innate, nor are they arbitrary. They are formed by historical circumstances. They can also change with historical circumstances.

What's in an Image?

Anyone growing up in American consumer culture understands the power of images. Images project messages about hopes and dreams, mien and demeanor, about who should be a scientist and what science is. What is the image of science? It is hard to identify one image that characterizes a typical scientist, yet the average American has at least a stereotypical notion of what to expect if introduced to "a scientist." In the same way that no woman identifies with the image of womanhood projected by Barbie, no scientist identifies fully with the popular image of science. Nonetheless, images cultivate a clientele. Do women see their futures reflected in the present face of science?

In 1957, about the same time Barbie was being designed, the well-known anthropologist Margaret Mead and her colleague Rhoda Métraux found that the average American high school student expected a scientist to be "a man who wears a white coat and works in a laboratory. He is elderly or middle aged and wears glasses . . . he may wear a beard . . . he may be unshaven and unkempt. He may be stooped and tired. He is surrounded by equipment: test tubes, Bunsen burners, flasks and bottles, a jungle gym of blown glass tubes and weird machines with dials." Students in Mead and Métraux's study further assumed that the scientist "is a genius" who creates new and better products for people. He has long years of expensive training and works long hours in the laboratory, "sometimes day and night, going without food and sleep." The students also thought that a scientist may have no other interests and "neglect his body for his mind." They were sure that "he neglects his family—pays no attention to his wife, never plays with his children. He has no social life . . . A scientist should not marry. No one wants to be such a scientist or to marry him." Note that the women, Mead and Métraux, conducting the study treated girls not as potential scientists but only as potential wives of scientists.[10]

Children persisted in conceiving of scientists as men well into the 1980s, when a collection of 165 drawings by secondary school children yielded the composite picture shown in Figure 3. Only two girls in the group drew a female scientist; none of the boys did. Even more strikingly, 82 percent of the student teachers imagined a scientist to be a man. In another study 86 percent of the girls (see Figure 4) and 99 percent of the boys described scientists as male—"a little on the crazy side with white hair that has not been combed in 40 years"; 1,580 out of 1,600 students imaged scientists as white.[11]

Both the general public and a good number of scientists themselves see science as both populated by men and identified with masculinity. Marcel LaFollette, in her study of the public image of American science from 1910 to 1955, also found an emphasis on physical stamina. The physicist Robert Millikan, who was the first to isolate the electron and measure its charge, was praised as "no ordinary man but ten men in one." So great was his dedication to science that he dismissed sleep as an activity for "ordinary" people. Men of Millikan's generation served long hours in the laboratory, where they built and used complex instruments. These same scientists, however, were often portrayed at home as baffled by can openers or potato peelers. Their image of absentmindedness and neglect of family was thought to prove their devotion to science.[12]

Crucial to this masculine image of genius has been transubstantiation of body into mind. Albert Einstein, a powerful icon of genius in the United States and abroad, evoked this theme of transcendence that stretches back at least to the seventeenth century: "I believe with Schopenhauer that one of the strongest motives that leads men away to art and science is a flight from everyday life with its painful crudity and hopeless emptiness, [a flight] from fetters of one's own ever shifting desires. A finely tempered nature longs to escape from personal life into the world of objective viewing and understanding . . . seeks in whatever way seems adequate to him to make a simplified and perspicuous picture of the world and thus to overcome the world of experience." Transcendence—the Platonic renunciation of body in favor of mind—is part of the ideology of modern rationalism. As formulated by Bertrand Russell in 1913, "the scientific attitude of mind involves a sweeping away of all other desires in the interests of the desire to know—it involves the suppression of hopes and fears,

Results of "Draw-A-Scientist" Test

$E=mc^2$

Eyeglasses
86%

Symbols of
Research
38%

Facial Hair
48%

Pencils/Pens
25%

Labcoat 63%

Male 92%   A SCIENTIST AT WORK

Figure 3. Results of a "draw-a-scientist" test. Most schoolchildren draw a man. Source: Kahle, "Images of Science."

loves and hates, and the whole subjective, emotional life." Great men of science are celebrated for ignoring bodily appetites: otherwise occupied, Newton is said to have often forgotten to eat the finely roasted chicken served to him in his study, and William Hamilton left half-empty plates accumulating for days as he worked.[13]

This form of scientific truth-telling rests on an unstated division of labor. The renunciation of everyday life often requires (though this is rarely acknowledged) that the scientist have someone—traditionally a wife, sister, mother, or housekeeper—to provide the necessities of life. Only a body without other bodies dependent on it can be truly transcendent. Descartes, we are told, had an illegitimate daughter, but he never had a pregnant body or a young child whom he fed, clothed, and held when she was sick

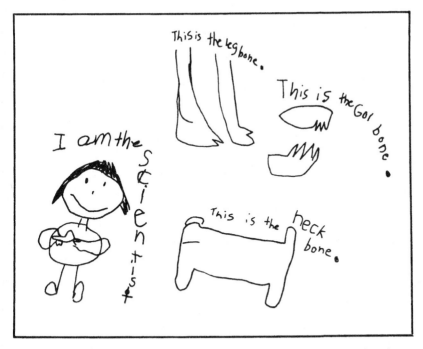

Figure 4. A first-grade girl's image of herself as a scientist. Source: Deborah Fort and Heather Varney, "How Students See Scientists: Mostly Male, Mostly White, and Mostly Benevolent," *Science and Children* (May 1989).

(he sent the child's mother away before the birth). Descartes's famous method required that he envision himself alone in his study, putting aside all previous learning, all beliefs, and all bodily needs.[14]

Even Marie Curie—a popular role model for scientific women—conformed to the image of a lonely, introspective scientist, clothed in a simple black dress and with severely pulled-back hair. In her youth, as a poor, driven student, she led a monastic life, becoming so lost in her studies in her cold room that she stopped bothering to light her stove or even eat. Physics so occupied her thoughts that she disdained even to learn how to make a broth. In our own day the astrophysicist Andrea Dupree recounts that early in her career she did not discuss good food, music, clothing, or traveling because doing so would have set her apart from many of her male colleagues who had "no social skills, or had them at an extremely low level." Only after she felt more secure in her profession did she reveal this range of her interests.[15]

While the perception of science as nonconformist potentially challenges both sexes, this complex of images—ranging from heroic masculinity to deep eccentricity—can create a barrier for women (though these images are also precisely the reason many women choose certain fields of science). The chemist Geri Richmond has described her gradual shedding of the trappings of traditional femininity to increase her credibility as a student of science. In high school Richmond was a cheerleader. In college she "dug into science and enjoyed it so much that I didn't come up for air until I graduated." To fit in with her male peers, she threw out her dresses, her pumps, her nail polish, and her makeup. She even jettisoned her hand lotion because she feared its fragrance might evoke her sex. And perhaps rightly so, for when Lise Meitner gave her first lecture at the University of Berlin on "The Significance of Radioactivity for Cosmic Processes" in 1922, the newspapers reported her topic as problems of "Cosmetic Processes."[16]

Not only is shedding the trappings of "femininity" often necessary for a woman to be taken seriously as a scientist, it is also often important for preventing unwanted attention to her sexuality. When Anne Kinney, a young astronomer at NASA's Space Telescope Science Institute, realized that her fashionable dress was a problem, she adopted what she calls her "chador"—jeans and a plaid shirt. Another woman describes herself and her female colleagues as "nuns in white lab coats," effacing their sense of fashion so as not to distract from the serious business of science. Claudia Henrion, in her fascinating book on the gendering of the culture of mathematics, reveals that women mathematicians "dress down" for work, changing into casual clothes when they reach the office and into dress clothes, and perhaps makeup, when they leave.[17]

While being feminine implies having at least one eye open to fashion, being a scientist requires a haughty indifference to appearance. (There are different dress codes for different disciplines: a longtime guard at the Princeton Institute for Advanced Study told me he could distinguish mathematicians from physicists, social scientists from art historians, from appearance alone.) And yet scientific women have been reproached for neglecting their femininity. James Watson wrote unkindly of Rosalind Franklin in his 1968 *Double Helix:* "By choice she did not emphasize her feminine qualities. Though her features were strong, she was not unattractive and might have been quite stunning had she taken even a mild interest in clothes. This she did not. There was never lipstick to contrast with her straight black hair and at the age of thirty-three her dresses showed all the imagination of English blue-stocking adolescents." At issue was not so much Franklin's appearance as her unbending refusal to be treated as an assistant in the lab rather than a researcher in her own right. For Watson, Franklin's appearance was emblematic of her insistent independence. A more "feminine" persona, he seemed to suggest, might have made her more submissive. Clearly Rosy (as she was called only in her absence) had to be put in her place, for, Watson assured his readers, "the best place for a feminist was in another person's lab."[18]

Watson, in not understanding that Franklin had adopted the sanctioned male inattentiveness to personal attractiveness, failed to see that women successful in traditionally masculine fields often assimilate or are assimilated to masculine codes of honor. The great German mathematician Emmy Noether was affectionately dubbed "der Noether" ("der" being a masculine pronoun), not only because "she was heavy of build and loud of voice" but also for "her power as a creative thinker who seemed to have broken through the barrier of sex." In an age now past, the highest compliment to a woman of science was to be made an honorary man. In 1908, when the British physicist Ernest Rutherford met the German physicist Lise Meitner for the first time, he exclaimed, "Oh, I thought you were a man!"—perhaps because no women except cleaners were allowed upstairs in the prestigious institute where she worked. Meitner developed her theories in a converted carpentry shop in the basement. Edwin Hubble similarly remarked that the distinguished astrophysicist Cecilia Payne-Gaposchkin was "the best man at Harvard."[19]

Women's entry into the white-collar workforce has required a revolution in clothing. The clothes of the 1950s often restrained women: straight skirts and spike heels made it impossible to walk for long distances; girdles made it difficult to sit comfortably for any length of time. Women's liberation in the 1960s and 1970s entailed adopting aspects of male dress and demeanor. In the 1960s women wore men's jeans made *not* to fit female proportions, often too large in the waist and too short in the crotch. Women even sculpted their bodies. Twiggy realized in female form the leanness of the young male. In the 1980s professional women adopted "execu-drag"—grey pinstriped suits with perky little red bow ties. Hoards of these women filled the streets of New York, wearing their running shoes to work and changing to heels in the office. At this juncture in the remaking of femininity, the MIT professor Vera Kistiakowsky's 1980 article on women in *Physics Today* ran opposite an instruments advertisement featuring a pretty woman whose highly glossed nails were intended to attract attention to the advertised products.[20]

Women are now more often able to find a wide range of practical clothing that is also attractive. I remember the freeing effects when, in the late 1980s, comfortable yet elegant flat shoes became available and acceptable to wear with a skirt. Women, and perhaps men too, are freer today to wear what they want. Feminists have also relaxed in their codes, which at one time dictated no high heels or makeup for women; a good feminist can now wear fire-engine red lipstick, if she so desires.

African Americans have similarly been required to assimilate to white images. Even today, black professionals—men and women—are subjected to close scrutiny. A reporter for *Newsweek* reflected on his conversation with a recruiter for the *New York Times:* "As we talked it became clear that he was focusing on such things as speech, manners, dress, and educational pedigree. He had in mind, apparently, a certain button-down sort, an intellectual, non-threatening, quiet-spoken type. That most whites at the *Times* fit no such stereotype seemed not to have occurred to him."[21]

The problematic portrayal of women in science parallels the problematic portrayal of women in the public sphere. The October 1993 issue of *Working Woman* ran a blank page to emphasize the absence of positive images of powerful women in the United States. Powerful women are often labeled unfeminine. Britain's first female prime minister was called "the iron lady," and this in a country that, unlike the United States, has a history of female heads of state stretching back to the audacious Elizabeth I in the sixteenth century.

Some women not only have denied their femininity in order to work as serious scientists but have obscured their sex entirely. As legend has it, Novella d'Andrea, who replaced her deceased father as professor of canon law at the University of Bologna in the fourteenth century, lectured from behind a curtain in order not to distract the male students by her great beauty. At the close of the eighteenth century the future prizewinning mathematician Sophie Germain followed courses at the newly opened Ecole Polytechnique in Paris (which, like most European universities, was closed to women at the time) under the pseudonym Antoine-August LeBlanc. The practice of feigning masculinity in order to enter the male world continued throughout the nineteenth century. The historian Kenneth Manning tells of a woman whose guardians sent her to the University of Edinburgh dressed as a boy. After taking a medical degree in 1812, "James" Barry joined the British army, and became the second-highest-ranking medical officer in the colonial military establishment. Her true sex was not discovered until after her death. In like manner, the first women to attend medical school in the United States in the 1850s, Elizabeth Blackwell, was counseled by a sympathetic professor to attend classes disguised as a man.[22]

Idealized images of scientists have not always been masculine. Throughout the seventeenth and eighteenth centuries science, knowledge, truth, and other abstract ideals were portrayed as women—as majestic as they were mythical. Mathematics is still sometimes referred to as "queen of the sciences."[23] These feminine images did not necessarily empower women. Women scientists of the day, such as the French physicist Emilie du Châtelet and the German astronomer Maria Cunitz, invoked them in ambivalent and diverse ways. Yet an idealized science in this earlier period was rendered female to act as mythical muse—a source of inspiration in the tradition of Boethius's Lady Philosophy or Dante's Beatrice—to real scientists who were mostly male.

Vestiges of these idealized feminine images remain. On the back of the Nobel Prize medals for chemistry and physics, designed in 1902, a female Natura holds a horn of plenty as Scientia (also female) lifts the veil from her face. Richard Feynman, a 1965 recipient of this medal, conjured up richly gendered imagery in his acceptance speech: "The idea [the space-time view of quantum electrodynamics] seemed so obvious to me and so elegant that I fell deeply in love with it. And, like falling in love with a woman, it is only possible if you do not know much about her, so you cannot see her faults." He went on to describe his prizewinning theory as "an old lady, who has very little that's attractive left in her, and the young today will not have their hearts pound when they look at her anymore. But, we can say the best that we can for any old woman, that she has been a very good mother and has given birth to some very good children." John Randall's classic history of philosophy from the same era also invokes a feminine past. In an extended allegory, philosophy as the parent of modern science is figured as a sensuous woman belonging "to the oldest profession in the world: she exists to give men pleasure."[24]

We today are feeling refreshing winds of change in images of science. Science magazines, recruiting brochures, and even textbooks are including more women's faces. But, even here, clumsy errors sometimes occur. The cover of *Science* magazine's 1993 issue on "Women in Science" featured a group of grade-school and adolescent girls (two of Asian descent, the rest European Americans—see Figure 9, p. 183). A critical viewer might question the easy portrayal of women as children: since the eighteenth century, women have often been considered men of incomplete growth or children of large stature, or otherwise identified with children. Some positive change has occurred in recent years, at least among the elites. A 1993 report showed that most Wellesley College undergraduates (still all women) no longer considered science and math "nerdy" or masculine fields. But 50 percent did believe that math and science required a "special calling" or genius. They also believed that one had to be "wedded" to science in order to be successful.[25]

Women in Professional Culture

Anyone who teaches American graduate students cannot help being struck by the silence of the women. Many women still feel stifled in classrooms, discussion groups, and professional meetings. Even today, after thirty years of the modern women's movement, many women in graduate school face a foreign culture. Betty Friedan remarked on the graduate women she met at Harvard University in 1983: "These women were awesome in their competence, but they made me uneasy. They seemed too neat, somehow, too controlled, constricted, almost subdued and slightly juiceless. The ambiance is so masculine, it alienated them somehow, though they might not be aware of it."[26]

We turn now from the images of science to its cultures, its internal workings, codes of honor, and unspoken rules. Many of the problems women face in science are common to other professions. Despite the fact that men and women of similar class and ethnic background grow up together and often establish intimate relationships, they live in separate cultures, each with its own styles of speech and nonverbal behaviors. In the 1970s Robin Lakoff identified what she called "women's language." Drawing on Japanese, in which women's and men's languages have distinct grammars, Lakoff began cataloguing the subtle distinctions in American language between men's and women's speech.[27] One striking thing about women is their silence in public places (in their own homes they are popularly portrayed as inexhaustible sources of chatter). Women have been muted for centuries by prescription, from the first century of the Christian era when Saint Paul taught that women, like children, should be seen but not heard, to the nineteenth century when medical doctors diagnosed publicly articulate women as hysterical. Today men and women speak with markedly different frequency in public settings.

A study of Australian archaeologists between 1988 and 1990 confirmed that at conferences men spoke publicly for longer periods than women (men averaged 32 seconds, with their comments ranging from 5 seconds to 4 minutes; women averaged 20 seconds, with their comments ranging from 5 seconds to 1.5 minutes). Women were more likely to ask questions, while men more often made comments and summary statements. Discussion also differed according to subject matter and audience composition. In sessions dealing with "soft science" topics, such as public archaeology or cultural resource management, women made up 60–70 percent of the audience and the frequency of male comments dropped to 31 percent. In sessions examining "harder" archaeological issues, such as physical evolution or Pleistocene archaeology, discussions were carried on primarily among men (88 percent). In the conference overall about two-thirds of audience questions and comments were made by men, and most of those by senior men.[28]

When women do speak, it is often with marked politeness. In order not to appear immodestly intelligent, forward, or pushy, women sometimes preface their remarks with apologies and disclaimers. A woman can be judged arrogant simply because she does not engage in what is considered appropriate womanly behavior—smiling, qualifying her statements, and tilting her head in a deferential fashion. Politeness also requires avoiding possible conflict by posing questions rather than making authoritative comments or issuing imperatives. Women, according to one study, are three times more likely than men to phrase directives as questions. This communication style makes women appear intellectually uncertain and hesitant.[29]

Women's lower status, politeness, and hesitations (feigned or real) invite interruption. Men tend to interrupt women in conversation more often than women interrupt men. As a result, women often speak quickly, feeling they should not impose on other people's time. Interruptions, of course, also follow status. Studies of faculty meetings show that speakers of higher rank tend to interrupt speakers of lower rank, even when all speakers are men.[30]

Women also tend to speak in conspicuously higher voices than men, a distinct liability in a culture that lends authority to the lower male voice (French women in particular cultivate an unnaturally high register). Even men strive, in certain instances, to lower their voices. A friend of mine at

Harvard Law School consciously lowered his voice when he joined the faculty, and a male radio announcer took up smoking cigarettes in order to lower his. Radio and television prefer to hire women announcers and newscasters with lower pitch. High, melodious voices are reserved for soothing young children; fathers may use a singsong falsetto when talking to their newborns.

Differences also mark men and women's nonverbal behaviors—facial expressions, gestures, touching, eye contact, use of space, and so forth. Women, expected to exude politeness in both speech and manner, are required, more than men, to smile. When listening, a woman may nod and smile to express attentiveness. If a woman does not smile, she may be perceived as angry. Women are overrepresented in professions that require smiling, such as nursing, teaching, daycare, serving as flight attendants or secretaries. According to the sociologist Arlie Hochschild, half of all working women, but only about a quarter of all working men, hold jobs demanding strenuous emotional labor.[31]

Men also tend to occupy more space in a room, beyond what differences in physical size might demand. Masculinity expands into available space—men cross their legs with foot on knee and stretch out along chair arms to mark out their territory. Femininity, by contrast, compresses the body in efforts to use as little space as possible. Women were traditionally taught to keep their legs crossed (at the knee or ankle) and to keep their elbows in.[32]

Men and women fall easily into expected and comfortable behaviors which may unintentionally perpetuate women's subordinate status. In a recent study of gender dynamics in laboratories, men and women were asked to divide a list of tasks. Men typically chose fewer "feminine" tasks for themselves when they believed that their partner was a woman than when they believed their partner was a man. Women also took on more feminine tasks if they believed their partners were men, even though they were not given information about the men's expectations. The authors of the study suggest that many well-intentioned people adapt to stereotyped expectations of colleagues without even being aware of doing so.[33]

Stereotypical expectations can also pervade other aspects of professional life. At work or meetings, men tend to talk to women about family, children, travel—anything but science. Andrea Dupree, former associate director of the Harvard-Smithsonian Center for Astrophysics, recalls that a male colleague, a member of the National Academy of Sciences, always spoke to her about an island where they both happen to vacation. At first she was flattered by the attention, but eventually she realized that while he spoke to men about astronomy, he spoke to her about the island and her holidays, not science. Now that she is a senior person in her field, she structures conversations with her colleagues so that they eventually turn to science.[34]

Leisure time does not necessarily lessen the uneasiness gender can interject into relationships between men and women. Even men and women who respect each other may experience a certain awkwardness during leisure-time conversation—time crucial for exchanging ideas and information and building a solid working relationship between colleagues. An easy rapport, vibrancy, and immediacy of exchange is often missing. Deborah Tannen has shown that, other things being equal (and there are many things determining whom we feel comfortable talking to—political proclivities, age, similar family situations, common backgrounds), male and female professionals prefer talking to persons of their own sex. Men's and women's discomfort with each other may have to do with the fact that men and women as groups tend to talk about different things. Men talking to men may discuss business, sports, public politics, and hobbies. Women talking to women discuss companions, friends, children, clothing, health, and perhaps also their situation as women in a particular profession.[35]

Similar gender asymmetries can influence student evaluations. A 1987 study revealed that when students thought they were evaluating a woman teacher, they gave her high marks if she nurtured them and devoted more time to them outside of class. The same was not true if the students thought they were evaluating a man. In practice, men who offer greater time and attention to students are not necessarily appreciated for this effort. Women faculty members are often expected to fit preconceived notions of proper female behavior, receiving poor evaluations if they do not smile, for example. Yet traditionally feminine behaviors conflict with students' expectations for professorial demeanor: students tend to rate stereotypically feminine women as less competent than women who present themselves in a more professional manner. Women can find themselves, once again, in a difficult situation: whatever behavior they adopt may be judged incongruous with the academic setting. Students have been known to challenge women faculty in ways they do not challenge men; some may find it difficult to accept a woman in a position of authority.[36]

The daily discomforts women experience in the professional world are often shared by men working in traditionally female fields, such as daycare or nursing. Do we look askance at male caregivers? Do we think: Is he a professional failure? Is he a child molester? Does he know how to hold a baby or nurture young children? Or do we shower him with welcome and reassurance—which merely serve to reinforce his outsider status? Even at the age of two, my son, while bleeding and being prepped for stitches, was sure that his nurse could not possibly have a mustache. The nurse handled the situation with the ease and grace symptomatic of long experience. How often has this man been asked about his choice of profession?

I have highlighted stereotypic behaviors of middle-class European-American men and women because little scholarship on other groups exists. Generalizations hold to some degree across ethnic groups, but not entirely. For example, European-American gender stereotypes turn upside down among Japanese: men are seen as cooperative and nurturing, women as individualist and competitive. Judged from a European or American point of view, Japanese men may seem to have more "feminine" leadership styles than European-American women. Asians as a group—both men and women—are often portrayed as holding cultural ideals that conflict with advancement in North American science. Asians are said to be "contemplative" or unassertive. One Asian-American woman, discussing the perplexities raised by stereotyping, remarked, "I've found the generalization 'girls can't do math' balanced nicely by the adage 'Asians are all math brains.' My presumed poor verbal skills as the child of immigrants were countered by the assumption that women are geniuses with words."[37]

It would be interesting to study the class backgrounds of minorities—men and women—who have made it in science. The historian of science Evelynn Hammonds has called attention to the intersections of class, race, and gender in her analysis of *The Great Tradition*. In this novel by Marjorie Hill Allee, Delinea Johnson, an African American (perhaps modeled after Roger Arliner Young, who completed her master's degree at the University of Chicago in 1926), is one of seven women doing graduate work in zoology at the University of Chicago in the 1930s. The other six are white. The European-American women live together, helping one another financially and with their studies. They regard Miss Johnson as dignified but distant. She does not join their group, nor is she invited to. At a crucial moment in the story the white women want to hire a cook and housecleaner. Delinea Johnson applies for the job, explaining that she must work to put herself through school. In the months that follow, she works as an equal alongside the white women by day, and by night she changes into her maid's uniform to prepare and serve their dinner. Delinea Johnson's domestic labor allows the white women to devote themselves to their studies. The advancement of professional women in the United States today often depends on the labors of poorly paid housecleaners and babysitters, many of whom are lower-class women or foreigners.[38]

While affirmative action has promoted the hiring of women and minorities in recent decades, these newcomers are often charged with having benefited from what some maliciously call "reverse discrimination." A graduate student at Stanford University says that often "it was pointed out to me that I was female and a minority and otherwise would not be at Stanford . . . Several times I nearly gave up because of this."[39] Newcomers to the academy are sometimes made to feel awkward as a means of keeping them in their place. African-American women report being "tokens," treated as symbols rather than individuals.

Black women, like other women, are also commonly overburdened with committee work and requests to give special attention to minority students. In the words of the president of Spelman College, Johnnetta Cole, black women are "the mammies of the academy": too often asked to assuage the fears of dominant groups, too often put in the position of making peace between diverse groups, and too often expected to comfort the weary and oppressed—demands that go well beyond the responsibilities of their formal positions.[40]

Competition, Science, and Sports

Nearly three decades ago the celebrated sociologist Robert Merton characterized science as working though "competitive cooperation." According to Merton, knowledge is gained through competition, but the products of competition are "communized" so the entire process strikes a balance between competition and cooperation. Today many women scientists characterize science as aggressively competitive and many of their male colleagues as discourteous and rude, pushing others aside in their efforts to be first. At a meeting to discuss how to bring more women into physics, held at the Aspen Center for Physics in 1994, the women urged their male colleagues to be more polite. One of their chief objections was something they called "macho-ness," which they defined as trying to prove oneself superior, being combative, and ignoring other people's ideas.[41]

Women's drastic underrepresentation in physics is commonly blamed on its highly competitive culture. Sharon Traweek, an ethnographer of high-energy physics communities, found that many physicists like to see themselves as independent, vigorously assertive, and competitive. As one male physicist put it, "only the blunt, bright bastards" make it. The physicist Heinz Pagels would not have disagreed: "A predominant feature in the conduct of scientific research is intellectual aggression, . . . a healthy sense of ego and intellectual intolerance." He claimed further that no great scientific discovery was ever made in a spirit of humility. This attitude extends beyond physics. The biologist James Watson admired Linus Pauling for his "unquenchable self-confidence" and claimed never to have seen his colleague Francis Crick (a former physicist) "in a modest mood."[42]

In this atmosphere, well-mannered, quiet young scientists of either sex are likely to have difficulty succeeding; and certainly women, subjected more than men to the cultural imperative to be modest, are put at a disadvantage. In the early twentieth century the influential Friedrich Hayek refused to admit women to his *Geistkreis,* one of Vienna's leading intellectual circles, because he considered it inappropriate to conduct free intellectual debate in their presence. He even refused to allow meetings to be held at the home of the man whose paper was to be discussed because he considered it uncivilized for a wife (wives were allowed to attend) to see her husband's work torn to pieces.[43]

Many women, of course, are competitive and aggressively so, and scientists are not alone in their strident behavior. Competition is endemic in North American professional life. Sociologists, however, tend to agree that the sciences, and particularly the physical sciences in America, require a sharp edge of competitiveness. European physicists have criticized their American colleagues for their overconfidence and abrasiveness. It may be significant in this regard that, as of 1990, the United States and South Korea had the lowest proportion of women physicists among the nations for which statistics are available (tied with about 3 percent each), while France and Italy, for example, both had between 15 and 25 percent. One Italian woman physicist reports that "in the U.S. I have to shout to get anyone to listen to me, and then I get accused of sounding hysterical."[44]

Competition in science is fostered by the "weeding out" process that students endure throughout their education. Students have been told: "Look to your right and to your left; only one of you will pass this course." This is said to be more characteristic of classes in the physical sciences than in the life sciences, but I experienced it as a graduate student in history at Harvard in the late 1970s. Women may fall victim to weeding-out practices more than men because competition intensifies their culturally induced sense of self-doubt. Women tend to take fewer risks, especially in fields where they are in a minority. Decreased risk-taking diminishes their opportunities to hone skills and develop an appropriate sense of confidence.[45] We should recall that the Enlightenment resolution to the "woman question" (the question of rights for women) included taking women out of competition with men. In the popular ideology of the eighteenth century, women were not the competitive equals of men but the delicate and pure "better halves" of their robust and assertive companions.

Excessive competition is one aspect of academic science that women report finding alienating; sports is another. Many academic departments arrange social events to bolster collegiality among faculty members and between faculty and graduate students. These gatherings often feature sporting events. The emphasis on sports again disadvantages women, this time outside the laboratory. Many women enjoy competitive sports and are athletic, and women's sports get more attention and respect than they used to. But men tend to be better at the sports most often chosen for such events. Women are once again faced with the need to succeed in typically male endeavors. No one suggests socializing around the balance beam or uneven parallel bars or while knitting or quilting.

One example of how sporting culture can shut women out will suffice. Susan Brantley, a Presidential Young Investigator (an honor bestowed upon the nation's top young scientists by the National Science Foundation), was hired in 1986 as the first tenure-track woman in the geology department at Pennsylvania State University. She got on well with her colleagues, even though she remained the only woman in the department for four or five years. One of the department's annual social events included showing the *Sports Illustrated* swimsuit video. The professor hosting the party showed Brantley the invitations before distributing them to the rest of his colleagues in an effort to make her feel welcome (or, as she remarked, "to ask my permission"). When Brantley indicated that she would not feel comfortable watching the video with her colleagues, he replied, "Well, you are a prude. You are ruining our fun." As Brantley looks back on it now, she says, "It was a small thing, but it made me feel uncomfortable to have to go to a full professor, someone who would sit on my tenure and promotion committee, and say that what he was planning struck me as very inappropriate. It exacerbated my feeling of isolation. I felt there was something wrong with me, that I was not really wanted in the situation—none of which is true because I generally feel very supported in my department."[46]

Men more often than women talk sports as a way of establishing contact with one another. They may believe (often correctly) that women do not understand or care about sports. Sports so influences academic discussion that common metaphors, even surrounding issues of gender equality, often come from this arena. We talk about creating a "level playing field," "fair play," going the "full nine yards," taking the initiative when "the ball is in your court," and so forth. An articulate and accomplished woman biologist told me: "I use these expressions at my peril, not knowing exactly what they mean."[47]

Sharon Traweek suggests that sports structures science in even deeper ways. The sports team, she argues, provides one influential model for working groups in American physics (Japanese physics, by contrast, is modeled on household structures). The group leader in American physics is like a coach directing a team of football players, each of whom has specialized skills. The coach, the only member with a view of the whole process, designs the team's strategies and tactics. The team survives as long as it continues to win.[48] The head of a department of marine ecology told me that one of the things she likes to know about graduate students (as a predictor of future success) is whether they have played team sports.

The call by women at the Aspen Center for Physics for their male colleagues to be more courteous recalls the struggles surrounding the creation of modern scientific styles and women's place in these debates. Scientists have developed certain ways of interacting and communicating both in person and in print. Their goal is to promote the growth of knowledge, and many believe that science is pushed forward most efficiently by competition between individuals and labs. Scientists write their articles in a terse passive voice; suppression of the subject buoys claims to objectivity. Many forget that science has a style, and that that style is a product of history.

Since its beginnings, modern science has been involved in a number of struggles over intellectual style. In the seventeenth century there was the struggle over the desired character of scientific language: should language retain the allegorical richness of the ancients, or adopt the more flat-footed precision of the moderns? In the eighteenth century scientists tried to cleanse "nature, the earth, the human soul, and the sciences of all poetry." The struggle, as Wolf Lepenies has described it, was one between science written in a literary or even poetic style and science written in dry and technical terms, with many tables and few well-chosen words.[49]

One important axis in these larger struggles was the issue of gender. The grand salons of seventeenth- and eighteenth-century Paris and other major European cities offered examples of intellectual institutions run by women (though not *for* women: the women of the salons served as patrons primarily to promising young men). Salons competed with universities and academies for recognition as institutions of learning, and can be seen as offering an alternative way of organizing intellectual life. The salons cultivated a distinct style of intellectual exchange: women allegedly brought to scholarship "a more varied vocabulary, greater nobility in diction, and more facility in expression." *Salonnières* (women who attended salons) contrasted the loathsome *pédant*, who pursued serious learning to the exclusion of social graces, with the *savant*, who combined knowledge with social refinement and eloquence with science. The women of the salons developed rules of etiquette requiring that a certain gentility govern energetic intellectual exchange. They also saw women as a crucial element in cultivating this politeness. As the eighteenth-century writer Madame Lambert put it, "Men who separate themselves from women lose politeness, softness, and that fine delicacy which is acquired only in the presence of women."[50]

We today can recognize that, as in the case of gendered images of science, there was no essential connection between the sex of the participant and these styles labeled masculine or feminine. The elaborate pageantry of gallant society emerged not from qualities innate to women but from the contours of aristocratic life. The "feminine" style, though couched in the language of gender, was an artifact of urban elite (often French) culture. As Madame Lambert described it, salon life joined the politeness and delicacy of the grand world of rank to the energy of intellectual work. In the salon, where advantages of rank outweighed disadvantages of sex, aristocratic women served as patrons to bourgeois men, showering the new rich with the *parfum de l'aristocratie*. The ethos of conviviality, as the historian Roger Hahn has pointed out, was common to elite aristocratic culture, both male and female, and only later became associated more closely with femininity.[51]

Salon women so influenced scholarly style in eighteenth-century Paris that the philosophe Jean-Jacques Rousseau launched a vicious counterattack. In the presence of women, Rousseau complained, men are required to "clothe reason in gallantry," to polish their conversation and be satisfied by jokes or compliments. Rousseau advocated a more vigorous form of scholarly exchange and deployed military metaphors to reinforce his point. Ideas, he argued, can be cultivated only on "the field of battle." In the absence of women, a man will feel himself attacked by all the forces of his adversary and will use "all his own force to defend himself." Only through this combative process, Rousseau believed, does the mind gain precision and vigor. The historian of science Martin Rudwick has noted the widespread use of military metaphors in scientific debates. The great Devonian controversy was likened to a "battle field," with "attacks and counterattacks," "frontal assaults," and so forth. Opponents pulled out their "heavy artillery," effectively bombarding and demolishing their intellectual enemies.[52]

By the late eighteenth century scientists and philosophers of science were championing a science stripped of metaphysics, poetry, and rhetorical embellishment. In Lavoisier's words, the language of science should restrict itself to "the series of facts which are the objects of the science, the ideas which represent these facts, and the words by which these ideas are expressed." By the middle of the nineteenth century the elimination of poetry from science had become normative and was said to be a natural stage in the evolution of human thought. In Claude Bernard's view, poetry was the first and most primitive of three stages of scholarship, succeeded by philosophy and finally science. Literature was banished from science under the disgraceful title of the "feminine." Equating the poetic with the feminine ratified the exclusion of women from science, but also set limits on the kind of language male scientists could use. While scientists strove to make facts transparent through what they considered unobtrusive language, certain sanctioned modes of expression merely replaced others.[53]

The critical nexus between the military, sports, and some fields of science persists today. The sociologists Bruno Latour and Steven Woolgar liken the laboratory to a "battalion headquarters at war." The biologist Richard Lewontin writes similarly that "science is a form of competitive and aggressive activity, a contest of man against man that provides knowledge as a side product. That side product is its only advantage over football." James Watson's bestseller about the discovery of DNA, *The Double Helix,* also bristles with battle metaphors. All of this, intentionally or not, tends to "sideline" women.[54]

There are, then, many aspects of scientific culture that tend to alienate women. Sociologists have studied the demographic characteristics of women who succeed in science (the economic and educational backgrounds of their parents, the schools they attended, the courses they took: see Chapter 3), but few have looked at the cultures of the sciences in which women are well represented. What can we learn from them? One science receiving attention in this regard is primatology, in which women receive 78 percent of the Ph.D.'s and are recognized as some of the leaders in the field. Work in primatology is not particularly glamorous. Primatologists, like other field researchers, go through long years of grueling training, often work in difficult climates and under challenging circumstances, and observe sometimes violent animals.

What has allowed for women's success? The primatologist Linda Fedigan suggests that primatology is a relatively young discipline, and historically women have fared better in new and rapidly growing disciplines that are somewhat marginal (as primatology was in its early years). Further, primatology is a life science allied with anthropology, psychology, and animal behavior—fields within which women have flourished. Within primatology, it is notable that women are more likely to work in social behavior than in anatomy, taxonomy, or physiology. Primatology also has strong female role models both in the popular press (Jane Goodall, Dian Fossey) and in academic circles (Jane Lancaster, Alison Richard). Finally, primatologists have cultivated an atmosphere welcoming to women. Men in the field—Louis Leakey, Sherwood Washburn, and others—have trained and supported women (sometimes perhaps for the wrong reasons) and the field has shown itself to be collegial and responsive to criticisms of sexist language and theories (see Chapter 7).[55] I might add that primatology has not been a "big science." Until recently there has been room in primatology for researchers working alone or in small, cooperative groups. As primatology becomes more and more dependent on large groups and long-term efforts, it will be interesting to see whether women will be the ones heading these efforts.

◖◕ 5

Science and Private Life

Perhaps the worst thing a professional woman can do is marry a professional man. For many men, marriage is a distinct advantage: married men with families on average earn more money, live longer, and progress faster in their careers than do single men. For a working woman, a family is a liability, extra baggage threatening to drag down her career. Though women still live longer than men, combining the incongruous responsibilities of work and family can be harmful to a woman's health. Working women with three or more children are at greater risk for heart disease than working women without children.[1]

Throughout the 1980s institutions sought to "level the playing field," to equalize initial conditions for men and women in the workplace. The projected playing field, however, was bounded by institutional walls. Few considered the inequalities still plaguing private lives. Science—like professional life in general—has been organized around the assumption that society need not reproduce itself, or that scientists are not among those involved in the day-to-day tasks of reproduction. While this may be true for many male scientists, it is not true for most female scientists. Professional women are still responsible for most domestic labor and child care. As the historian Gerda Lerner has written, "The sexual division of labor which has allotted to women the major responsibility for domestic services and the nurturance of children has freed men from the cumbersome details of daily survival activities, while it disproportionately has burdened women with them."[2]

Can a woman who takes charge of domestic life compete professionally with a man or woman who does not? As women have begun taking their place in the professions, certain aspects of professional life have been reformed. The domestic sphere, however, has never been subject to affirmative action or legislative amendments requiring redistribution of domestic labor. Women in heterosexual relationships generally remain—reluctantly or not—in charge of hearth and home. Consequently, women who go out to work add a demanding profession to what used to be considered a full-time job. Being a scientist and a wife and mother is a burden in a society that expects women more often than men to put family ahead of career.

Domestic arrangements *are* part of the culture of science. Despite the historical distinction between the domestic and public spheres, private life is not separate from public life. Nor is the conflict that many women encounter between family and career just a private matter. Professional culture has been structured on the assumption that a professional has a stay-at-home wife and benefits from her unpaid labor.

Do women actually do more of the domestic work in heterosexual households where both partners are professionals?[3] As one might expect, there is some disagreement between men and women on this score. About 43 percent of men say they share child care equally with their spouses, but only 19 percent of women agree. A 1993 study by the Families and Work Institute of New York found that in dual-earner families women did 81 percent of the cooking, 78 percent of the cleaning, 87 percent of the family shopping, and 63 percent of the bill paying. Men outperformed women only in household repairs (taking responsibility about 91 percent of the time). This study did not include yard or car care, for which men may take significant responsibility. Professional women work roughly fifteen hours longer at home each week than do their husbands. Over a year this adds up to an extra month of 24-hour days. And they sleep less. Married women tend to get 20 minutes less sleep per night than their partners; women with children get 40 minutes less sleep per night than their husbands. In a week the average working mother gets 4.6 hours less sleep than the average working father. This adds up to almost 10 days less sleep per year. These patterns hold even in families in which the woman significantly out-earns the man.[4]

Child care arrangements, like any other aspect of culture, are not etched in nature but configured by social contingencies and political priorities. Child care was less of a burden for upper-class women of the eighteenth century than it is for professional women today. In the eighteenth century children from wealthy urban families were handed over minutes after birth to wet nurses and reared in the countryside. Parents might not see a child again until it was seven years old—about the time upper-class boys were sent away to boarding school and girls were given over to governesses. Modern motherhood—the notion that the woman who bears a child should also bear primary responsibility for the care of that child—took on a new cultural force in the late eighteenth century, when women were encouraged to return to the home and care for their children.[5]

The professionalization of modern science in the eighteenth and nineteenth centuries occurred in step with the new value placed on mothering. Mothers have been made to feel terribly guilty if they "neglect their children" by working outside the home. These attitudes have not changed a great deal in recent years. A 1993 survey of women undergraduates at Wellesley College revealed that 90 percent believed women with infants should not work full time. Half of these students also believed that fathers with infants should not work full time.[6] (It is unclear, if neither the mother nor the father is working, who is bringing home a paycheck.) Professional women find themselves at an impasse: the years from ages 22 to 40, crucial for establishing a successful career, are also prime childbearing years.

Women themselves—to say nothing of childbearing or childrearing—have long been considered disruptive of serious scientific endeavor. Again, these attitudes are deeply rooted in the past. Ancient Hebrew tradition held that through contact with women men lost the power of prophecy. In the Middle Ages the life of the mind was a celibate one. Intellectual life took place in monasteries, and monasteries influenced their successors—universities.[7] Professors at Oxford and Cambridge, for example, were not allowed to marry; late into the nineteenth century celibacy was still required. Not long ago there was a historian of science at Harvard who offered the following prescription for great science: be a genius, get little sleep, and have no sex.

Until the early twentieth century U.S. women's colleges required their women faculty members to remain single, on the grounds that a woman could not carry on two full-time professions at once. Male faculty members at the same colleges were, by contrast, required to marry, presumably to neutralize their potential danger to the students. Only Bryn Mawr College hired single men.[8]

In the 1970s and early 1980s many academic women, including scientists, avoided having children. In this context Jonathan Cole and Harriet Zuckerman published their highly counterintuitive finding that marriage, and even the successive births of children, did *not* impede a woman's scientific productivity. Astonishingly, they found that married women with children published as many papers each year, on average, as single women.[9] Despite their good intentions, Cole and Zuckerman advocated nothing less than the "Superwoman"—the highly organized, efficient, professional woman who was also a loving wife and a perfect mother—the woman who "could have it all" and who could do it all.

After the antichild decade of the 1970s, professional women began having families, but often in secretive ways. Women tried to "hide" pregnancies as long as possible. I had both of my children during research (not maternity) leaves so that my colleagues would never see me pregnant. Women, sometimes pretending even to themselves that they were not pregnant, refused to slow down. The chemist Geri Richmond recalls her first pregnancy: "I was sick every day for seven months, but I just kept going. I just didn't want people to look at me as a feminine creature." Women even "scheduled" babies. The physicist Ellen Williams timed one pregnancy so that she could have the baby during a sabbatical and took all her sick leave and vacation time to bear the other. The biologist Deborah Spector had labor induced on a three-day weekend so she could attend a student's thesis defense the following Monday. The Columbia University physicist Elena Aprile taught throughout her second pregnancy, and one month after the baby was born submitted a major research proposal to develop a gamma ray telescope for NASA.[10]

The goal for these women was to have babies without maternity leave, without a pause in productivity, without appearing to be different from their male colleagues. The result was that they did it at a high cost to themselves and their partners within institutions structured to suppress such things. Women report that they continue to produce scientific papers at the expected rate by eliminating almost everything but work and family. What went first was time for themselves—movies, novels, workouts, dinner parties. They also lost the flexibility to stay late at the lab or to engage colleagues in informal discussions.

Though professional women now more often choose to marry and have children, they are still less free to do so than many of their male colleagues. While 94 percent of male scientists in the United States are married, only 70 percent of female scientists are. The number of unmarried women is higher in certain groups: 38 percent of women chemists are single compared to 18 percent of the men, and African-American women scientists as a group are unlikely to marry.[11] A larger proportion of female than male scientists remain childless: 37 percent of women scientists over the age of fifty compared with only 9 percent of the men. Again, the proportions are magnified in some areas: only 17 percent of the women who are full professors of engineering have children, while 82 percent of the men do.

Women are becoming more aggressive about wanting the same options as men. When a series of articles in *Science* in 1994 seemed to suggest that women who intend to succeed in scientific careers should "forget about babies," seventy-seven women signed a letter of protest. Traditional attitudes, however, still reign in some European countries, such as Germany, where a neurobiologist at Tübingen's developmental biology institute has reported that she knows a dozen young women scientists who have had abortions because they thought that having a baby would end their careers.[12]

The assumption that a professional had a stay-at-home wife manifested itself in the "family" wage that men used to be paid. (In the same era, a woman's wages were typically considered supplemental and not essential to her survival even when she was single, divorced, or widowed.) Though men's salaries are still on average higher than women's, they are no longer justified as family wages.

While many professional men are still married to homemakers, nearly every married woman professional comes with a professional man. A stay-at-home husband is a rare luxury. Women in Europe and North America by and large practice what is known as *hypergamy,* the tendency to marry men of higher (or at least not lower) status than their own. Consequently, more professional women than professional men are married to professionals. While only 7 percent of the members of the American Physical Society are women, 44 percent of them are married to other physicists. An additional 25 percent are married to other types of scientists. A remarkable 80 percent of women mathematicians and 33 percent of women chemists practice disciplinary endogamy.[13]

These statistics do not tell the whole story. If he has arranged it well, the man with a traditional division of domestic labor works hard all day, but returns home to enjoy hot food, a well-organized family, and a fully orchestrated social life. Two people are working to produce one professional (and, of course, only one paycheck). A dual-career couple with young children, by contrast, cannot fall into relaxation mode at five o'clock but must face family responsibilities. Parents stop at the soccer field, the dance studio, the after-school program to pick up tired and often fussy children. They arrive home to an empty house with little food in the refrigerator. Shopping at a market overcrowded with harried working men and women comes next. One parent then watches young children or supervises homework while the other cooks dinner. Someone cleans up. For both parents, the need to plan the day around a five-o'clock pickup takes its toll on such things as spontaneous discussion with colleagues.[14]

There is often more to it than that. Behind many a senior professional man is a helpful wife who does not necessarily have her own full-time position. Some traditional stay-at-home wives also serve as research assistants, book editors, and discussion partners who dedicate long hours of service to their husbands' careers. The labors of these wifely assistants are for the most part invisible. Productivity studies do not take into account the untold hours and talents many wives contribute to their husbands' careers. "For many centuries the talents of women were directed," writes Gerda Lerner, "not toward self-development but toward realizing themselves through their husbands' careers. Women have . . . nurtured [men] in a way that allowed men of talent a fuller development and a more intensive degree of specialization than women have ever had."[15]

Some dual-career couples share interests, some even collaborate, but their professional work must justify two careers, not one. Cole and Zuckerman have argued that marriage to a man in her field is a distinct advantage for a woman scientist. Women scientists married to other scientists publish, on average, 40 percent more than women married to men in nonscientific fields, even more if the husband is more established professionally than the wife. Presumably the woman's productivity is enhanced by access to her husband's professional networks. Perhaps because some women have taken advantage of their husbands' contacts (this certainly was more characteristic of earlier generations than it is today), when couples work together it is sometimes assumed that the important conceptual work has been done by the man. This problem can be of such proportions that women stop collaborating with their spouses. The problem of women receiving due credit for their work is an old one: Emilie du Châtelet, an eighteenth-century scientist, was better known for her liaison with Voltaire than for her physics. As one of her contemporaries remarked, "Women are . . . like conquered nations . . . whatever originality, greatness, and sometimes genius they possess is considered only as a reflection of the spirit of the famous man they loved."[16]

Dual-career couples working in very different fields are subject to the stresses and strains of divergent careers. Traditional families can set their rhythm according to the ups and downs of one career. There are deadlines, periods of celebration and recuperation. With two careers, companions are rarely in sync. One may have just finished a project and be ready to relax while the other races toward a deadline.

The woman in a dual-career couple is often saddled with the burden of a "second shift" in addition to all the stresses of her profession. Men are beginning to take increased responsibility for children. Indeed, an increase in this area would not be difficult, considering the brevity of the time fathers spent with their children a generation ago. A study in 1971 reported that fathers spent an average of only 37.7 seconds each day communicating with their babies during the first three months of life. More recent studies show that 70 percent of working women take primary responsibility for their children, compared with 5 percent of working men. On workdays fathers spend an average of two and a half hours with their children, while mothers spend more than three and a half hours with them. While fathers "help out" with specific tasks, few assume regular responsibility for child care.[17]

Even women who pay others to do household tasks typically hire and train the employees and oversee their work. It is wrong to imagine that these paid laborers, no matter how dedicated, can replace the traditional wife and mother. Professional women today run households as women in the Middle Ages ran manors, overseeing and coordinating domestic labor. In addition, many shoulder the emotional burden of worrying about whether their children are properly cared for. Even in families where the woman is the main breadwinner, men do not share household responsibilities equally. A recent study reported that although 85 percent of women senior executives earn more than their spouses, 49 percent still have primary responsibility for household tasks, and 56 percent have primary responsibility for child care.[18]

Many fathers are beginning to place a greater premium on parenting; they spend more time with their children and say they understand their children well. The sad part is that fathers who share family responsibilities may be disadvantaged in today's professional world. These men are beginning to make the same kind of trade-offs that women traditionally made, turning down promotions or jobs that require extensive travel and overtime. There is now what has been identified as a "daddy penalty," whereby male managers with professional wives earn some 25 percent less than those with stay-at-home wives. A 1995 survey showed that managerial men in large corporations, even young men with children, are disproportionately likely to have traditional families in which the wife stays home and manages the household.[19]

Dual-career couples also suffer from decreased job mobility. Moving from job to job can be crucial early in one's career; mobility helps one gather experience, find the right position, and improve one's salary and working conditions. Here, too, women have been more constrained. Because their husbands are often older and more established, wives have tended to follow their husbands, or given preference to their husband's career development.[20] It is rare that a man will follow a woman to a job.

In addition to shouldering weighty domestic responsibilities, professional women in our culture are subjected to a number of psychological and emotional burdens. In the 1960s physicians incorrectly claimed that women who did not have their children before the age of thirty had a greater chance of developing endometriosis. Today the medical community haunts women who delay childbirth (often for professional reasons) with the specter of an increased likelihood of breast cancer or birth defects. Professional mothers are also sometimes blamed for an astonishing variety of the ills of modern society. In an infamous case, the *Canadian Journal of Physics* published an article (in an issue devoted to the kinetics of nonhomogeneous processes) blaming working mothers for everything from increased student cheating to drug use, insider trading, infidelity, embezzlement, teenage sex, and corrupt political practices. In an argument worthy of the eighteenth-century proponent of motherhood Jean-Jacques Rousseau, the University of Alberta chemist Gordon Freeman argued that women are "equipped by nature to be nurturers" and that the children of working mothers suffer "serious psychological damage." What is shocking is that this article made it through peer review. In response to the ensuing uproar, the editor of the *CJP* called the whole affair a "most interesting and complex mixture of scientific publishing, political correctness, vulgar politics of protest, . . . media manipulation, and government agency damage control."[21]

Women considering careers in science cite the difficulties of combining career and family as a major concern. In the mid-1960s Alice Rossi asked female college graduates why so few intended to go into science or engineering. Among the reasons given were the difficulty of combining work and family (54 percent), the desire to work part time (38 percent), the unfeminine image of the woman scientist (23 percent), and inadequate skills (6 percent). A similar survey conducted in 1991 found that the difficulties of combining a career in science and family responsibilities still rated highly (24 percent of the respondents).[22] In a survey of Stanford graduate students in science and medicine, nearly two-thirds of the women reported expecting or experiencing difficulty integrating work and family, while only a third of the men felt this way. Gerhard Sonnert and Gerald Holton, in their study of high achievers in science, found that the two obstacles women mentioned most often were family demands and their spouses' careers; almost none of the men cited either obstacle.[23]

The "playing field" of science will never be leveled as long as child care and household management continue to be seen as primarily a woman's concern. It is no more true that men with Harvard Ph.D.'s are genetically incapable of doing laundry than that women are genetically incapable of doing math (it is revealing, though, that the former is less studied than the latter). Men will have to learn to pull their weight at home; women, who often control domestic space, will have to learn to share that control (in other words, men may have their own ways of organizing and per-forming domestic tasks—we cannot impose our ways on them). Partners need to agree on a division of domestic labor that puts the man in charge of half the work and allows him to assume half of the responsibility. It is not enough for men to "help out"; they must take responsibility for the physical, intellectual, and emotional functioning of family life. Women may initially have to "mentor" men to set them on the road to indepen-dence in domestic thinking.

Professionals currently work within social arrangements hammered out in the eighteenth century, when a professional was considered a self-standing individual but was, in fact, the male head of a household. Women scientists who, more often than men scientists, are married to other professionals do not easily fit this mold. In order to bring women into science, we need to restructure the professional and domestic worlds.

For a long time anti-nepotism rules dealt with intellectual wives: wives simply were not hired by institutions where their husbands worked. The Nobel Prize–winning physicist Maria Goeppert Mayer experienced these policies at first hand. She told younger women interested in physics in the 1950s that "it was hard to be a woman physicist" but "nearly impossible to be a married woman physicist."[24] Anti-nepotism was challenged in the 1960s, and hiring of couples has become quite common and often nec-essary to retain good faculty. The practice is still fraught with difficulty, however. It is still illegal, for example, for public employers to ask about a person's family in an official interview. These laws were put into place primarily to protect women, the rationale being that family is supposed to be a private matter and not of any concern to the employer. According to this way of thinking, individuals should be considered on merit alone.

Personal considerations (whether or not a potential employee will be commuting, will be seeking parental leave, and so on) are not supposed to be taken into consideration.

How realistic are these practices today? Universities, government, and industry stand to lose if their employees commute from town to town and even across continents.[25] People can do only so many things: they can commute, teach, and have children; or they can teach, do research, and have children, but they cannot commute, do research, teach, do a full load of committee work, and also enjoy a vital family life. There are limits. Today we should perhaps ask about family situations, especially when discussing employment for women, since, as we have seen, the phenomenon of the dual-career couple is more closely associated with women's than men's employment. If employers do begin to consider people as cooperative units rather than as supposed individuals, what will change? Spousal hiring (more appropriately "partner hiring" to include unmarried or same-sex couples), the hiring of both members of a couple, is now a widely accepted practice but has its own difficulties. Despite a partner's qualifications, he or she may not have the specialty a department or unit is seeking. If units are instructed by deans to hire partners, what happens to academic freedom? How low on the quality pole will a department or a firm go to hire a partner?

The current state of chaos in personal lives and institutional policies indicates a need for restructuring the relationship between professional and domestic lives in the twentieth-first century. One proposal has been that the National Science Foundation should initiate an employment program for couples, providing funding for the partner of a scientist (male or female) for six years, after which time the institution would evaluate the partner for tenure and, if successful, provide permanent employment.[26] Another suggestion has been for institutions to maintain a number of unallocated positions to be used to hire highly qualified partners; this would be similar to the University of California's program for hiring outstanding women and minorities specializing in areas not targeted in departmental strategic plans. While creating new positions for partners may have very positive effects for all concerned, it may also lead to the perception—as in the case of affirmative action—that a partner is not well qualified. An institution's willingness to create a new position often depends more on how badly it wants the primary candidate than on the qualifications of the accompanying partner.

Institutions should also look favorably on solutions suggested by potential employees. In 1976 Jane Lubchenco and Bruce Menge, two highly qualified marine ecologists, split a single assistant professor position in the department of zoology at Oregon State University into two half-time positions.[27] (They had been on the tenure track at Harvard and the University of Massachusetts in Boston, but wanted half-time positions when they had children.) These part-time but mainstream appointments allowed them to spend more time with their young children without sacrificing their teaching and research. They were subsequently both tenured and became full-time professors, and Lubchenco won a MacArthur "genius award." This arrangement required a supportive faculty, university administrators willing to make and sustain unconventional arrangements, and an institution allowing part-time positions to be tenured. Positions need not, of course, be split only between the members of a couple. Unrelated persons should also be able to take what Lubchenco and Menge called "fractional but mainstream" positions.

While it is important that institutions support employees' solutions to structural problems, half-time positions are not a viable solution for many couples who need or want two incomes. Even two academic salaries—especially early in the partners' careers—do not make it easy to sustain a family in Manhattan or some other large cities. Lubchenco and Menge split a position at Oregon State, where the cost of living is relatively low, and even they experienced "fiscal austerity."

Both Lubchenco and Menge also complained that they ended up working considerably more than half time, that neither was eligible for university awards earmarked for full-time faculty members, and that they were constantly being compared to each other. Shared positions are, in fact, not very common, and pose real problems. Administrators fear that the couple may get a divorce, for example, or vote the same way on departmental issues.

Adjusting the tenure clock is another solution sometimes offered by universities for faculty members who need time off for family matters. This option tends, however, to sidetrack women's careers (these programs, though available to both men and women, are most often used by women). Even if the tenure clock is stopped inside the institution, colleagues outside the university and granting agencies may not take this into consideration when asked for evaluations.[28] Another proposal of dubious merit is that institutions provide "soft money" research positions for partners of tenure-track faculty members. This produces a ghetto for women: three-quarters of partners on soft money are women. Women in these positions feel they have been denied real jobs.

Family responsibilities are not the only reason people may seek flexibility in employment. Creative people tend to be multitalented and usually want time to enjoy music, the arts, sports, or politics. Mythology to the contrary, science is not an eighteen-hour-a-day job. No one can sustain health, let alone creativity, under such circumstances. Moreover, people's best ideas sometimes come while they are relaxing. One thinks of Kekule's benzene rings dancing in the fire, Heisenberg and Bohr's peripatetic discussions of quantum mechanics and relativity, and Aristotle's maxim that theory is a luxury of leisure. A certain intensity is needed to succeed in creative endeavors, but we should avoid misplaced shows of stamina that tend to exclude people with rich and varied interests.

III

GENDER IN THE SUBSTANCE OF SCIENCE

The famed australopithecine "Lucy" was very possibly male, yet by convention all smallish skeletal remains are female—which obviously skews the interpretation of home-sites and grave-goods.

ADRIENNE ZIHLMAN, physical anthropologist, 1997

If we can identify the role of human agency in the making of knowledge, we—as women and as scientists—could know other things in new ways.

JOAN GERO, archaeologist, 1993

What was decided among the prehistoric protozoa cannot be annulled by Act of Parliament.

SIR PATRICK GEDDES and J. ARTHUR THOMSON, biologists, 1889

6

Medicine

MANY people may be willing to concede that women have not been given a fair shake, that social attitudes and scientific institutions are in need of reform. They may also be willing to concede that women are excluded in subtle and often invisible ways. They stop short, however, when it comes to analyzing the effects of gendered practices and ideologies on knowledge. Does the exclusion of women from the sciences have consequences for the content of science?

Since the Enlightenment science has stirred hearts and minds with its promise of a "neutral" and privileged vantage point, above the rough and tumble of political life. Men and women alike have responded to the lure of science: "the promise of touching the world at its innermost being, a touching made possible by the power of pure thought."[1] The power of Western science—its methodology and epistemology—is celebrated for producing objective and universal knowledge that transcends cultural restraints. With respect to gender, race, and much else, however, science is not value-neutral. Gender inequalities, built into the institutions of science, have influenced the knowledge issuing from those institutions.

When aspects of science are sexist, scientists themselves may be unaware of it. There is no evidence, for instance, that the great eighteenth-century Swedish naturalist Carl Linnaeus intentionally chose a gender-charged term when he named a class of animals *Mammalia*. He may have done so naively, but he did not do so arbitrarily. As we shall see, his innovation responded to the world of human interests, political tensions, and common assumptions in which he lived. We are beginning to appreciate more and more the contingencies of scientific knowledge, and especially what is forgone in the choice of one particular course of research rather than another.

In the field of medicine, the founding in 1990 of the National Institutes of Health (NIH) Office of Research on Women's Health and the Women's Health Initiative of 1991 were a triumph for feminism. Between 1990 and 1994 the U.S. Congress enacted no fewer than twenty-five pieces of legislation to improve the health of American women, ranging from a requirement that women be included in clinical trials to new federal regulations for mammography. Taking women's health seriously did not require new technical breakthroughs or simply more women doctors— though those changes did help. Nor did it emerge from the supposedly self-correcting mechanisms of science. As Bernadine Healy, a former director of NIH, commented, "Research alone cannot correct the disparities, inequities, or insensitivities of the health care system." Reforming aspects of medical research required new judgments about social worth and a new political will.[2] *There needs to more than laws to to change the mind set that changes the science and women*

History

Today, in the heat of the women's health movement, grave concerns have been raised about inadequate knowledge of the female body. Contrary to popular belief, however, Western culture has committed enormous resources to the science of woman, studying the physical, moral, and intellectual character of "the sex," as women once were called. But much of this research was not intended to contribute to women's health and well-being. "Sexual science"—the intense scrutiny of sexual difference—played an important role in attempts to resolve debates about women's proper role in society and the professions.[3]

In 1543 Andreas Vesalius, the celebrated father of modern anatomy, prepared two paper dolls or manikins designed to be cut out and "dressed" with their organs to teach medical students the position and relation of the various viscera. One doll represented a female figure and displayed the system of nerves; the other represented a male figure and showed the muscles. Vesalius presented both male and female in order to demonstrate the position and nature of the organs of generation. When discussing parts of anatomy not having to do with reproduction, he did not differentiate male from female. In his instructions he stated: "The sheet [of organs to be attached to the male person] differs in no way from that containing the figures to be joined to [the female] except for the organs of generation."[4] In his presentation of male and female bodies, Vesalius set one pattern that has persisted to the present: men and women's bodies are considered biologically interchangeable, except for those parts directly relating to reproduction.

Vesalius's lack of recognition of nonreproductive differences between the sexes did not derive from an ignorance of the female body. From as early as the fourteenth century women had been dissected. The *Montpellier Codex* of 1363 includes an illustration showing the dissection of a female body, and the 1442 statutes of the University of Bologna reveal that the university received one male and one female body for dissection every year. A statute enacted in France in 1560 required midwives to attend the dissection of female bodies so they would be better able to testify in abortion cases. Vesalius himself based his drawings of female reproductive organs on dissections of at least nine female bodies. Hearing that the mistress of a certain monk had died, Vesalius and his helpers snatched her body from its tomb.[5]

For Vesalius, living in sixteenth-century Venice, seeing sexual differences in human bodies as limited to the sex organs was perhaps not so surprising. As a physician, he seldom treated male patients (that was done by lower-class barber-surgeons); as a man, he rarely treated female patients (that was the provenance of midwives). Classical divisions between medicine, surgery, and midwifery had long made midwives "specialists" in women's health.

We do not know what model of sexual difference informed midwives' practices. Midwives practiced medicine; they seldom wrote about it. Since they treated primarily women, they may not have developed theories concerning differences of sex. We do know that when women's health care was taken over by professional medicine in the eighteenth and nineteenth centuries, the experience of giving birth changed dramatically for women.[6] Without romanticizing midwives, it is possible to highlight certain differences between their practices and those of the man-midwives and their successors, the obstetricians. Whereas, for example, early modern midwives had assisted the mother not only with the birth but with other aspects of her daily life (by, for example, cooking and caring for the other children while the mother recuperated), man-midwives attended to the mother only during the hours of labor and eventually required women to give birth in hospitals—a process that removed the women from their support systems.

The demise of traditional midwives in early modern Europe had other consequences for women's health and well-being. Notably, women lost control over their fertility. As late as 1600, women in parts of Europe commonly had access to some 200 contraceptives and abortifacients, of both a vegetable and a mechanical nature.[7] Within Europe the decline of midwifery undermined traditional knowledge of contraception—knowledge that had typically passed along women's networks from mother to daughter, midwife to neighbor. As a result, European women of the nineteenth century had more children than their grandmothers did and understood less about their own bodies. the tea.

I do not mean to suggest that women need be looked after by women health professionals. I am advocating neither Victorian modesty nor cultural essentialism that teaches that women can better treat members of their own sex. The practices of early modern midwives were not always helpful to women. Midwives were often employed by the church or local governments to regulate illegitimate births, even at times forcing the woman during the pains of birth to reveal the father's name. In our century the Nazis revived the art of midwifery to benefit the "master race," not necessarily to benefit women. I am suggesting that until well into the eighteenth century women enjoyed care by experts in women's health who were neither academic physicians nor barber-surgeons. Female practitioners were not accepted as co-professionals in the nineteenth century, but were excluded from medical schools. As a result, women became more and more dependent on university-trained physicians for their care.

Professors of medicine such as Vesalius, then, could ignore nonreproductive sexual differences because women's health lay largely outside their jurisdiction. Early modern academic medical men could also ignore such differences because they had inherited an account of the distinctions between male and female, an account that was not challenged until the eighteenth century. From Aristotle's pronouncements on women as cold and wet to Darwin's notion of woman as a man whose evolution has been arrested, academics have viewed woman as an incomplete or lesser version of man, a "departure from type," a "monstrosity," or an "error of nature." Woman's tragic flaw, according to Aristotle, was her lack of vital heat to cook the blood and purify the soul. This lack of heat accounted for woman's weaker reason.[8]

The notion of woman as an incomplete or imperfect man—a deviation from the norm—has served as a cornerstone of Western views of sexual difference. Galen, the second-century Greek physician, popularized the idea that even women's sexual organs are but a lesser version of men's.

Galen taught that a woman has a "spermatic vessel" or penis similar to a man's, except that it is inverted and internal. As proof that women are merely incomplete men, Galen and Pliny recounted stories of women spontaneously transforming into men; for the most part this physiological inconvenience occurred on their wedding day. There was the woman in the time of Pope Alexander VI who, on the day of her marriage, "had suddenly a virile member grown out of her body." There was also the man at Auscis in Vasconia, now, at age sixty, strong, grey and hairy, who had been a woman until "at age fifteen, by accident of a fall, the Ligaments being broken, her privities came outward, and she changed her sex."[9] The transformation was not reversible, however. Galen argued that, though a woman might become a man, a man could not become a woman. The reason: nature strives always for perfection.

An important challenge to these notions emerged in the seventeenth and eighteenth centuries, coinciding with the formalization of women's exclusion from science. At the birth of modern science, the noble networks and artisan workshops gave women (limited) access to science (see Chapter 1). Women's incursion into serious intellectual endeavor was supported ideologically by the Cartesian wedge driven between mind and body, which fostered the notion that "the mind has no sex." The subsequent exclusion of women from science and public life required new justifications. Within the framework of Enlightenment thought, an appeal to natural rights could be countered only by proof of natural inequalities. An individual's place in the *polis* increasingly depended on his or her property holdings, and also on sexual and racial characteristics. Science, with its promise of a neutral and privileged viewpoint, came to mediate between the laws of "nature" and the laws of legislatures. For many, scientists did not have to take a stand in questions of social equalities because the body—stripped as clean of history and culture as it was of clothes and often of skin—"spoke for itself."[10]

The eighteenth century witnessed a revolution in sexual science. At this time academic medical men ceased to see the female body as a lesser version of the male and emphasized instead a model of radical difference. Sexual difference was no longer a matter of genitalia but involved every fiber of the body. By the 1790s European anatomists presented the male body and the female body as each having a distinct telos—physical and intellectual strength for the man, motherhood for the woman. In this context the first drawings of distinctively female skeletons appeared in Europe. Although these were drawn from nature with painstaking exactitude, great debate erupted over the distinctive features of the female skeleton. Political circumstances called immediate attention to depictions of the skull as a measure of intelligence and the pelvis as a measure of womanliness.[11]

The revolution in sexual science brought with it a new appreciation of woman's unique sexual character. Few physicians, however, were concerned about the implications of difference for health care. For the most part, academic study of sexual differences was designed to keep women in their place. The eighteenth century reestablished on new foundations the view that "biology is destiny": that the failure of women to create great science was to be found in their "nature." The stage was set for the virulent sexism of the nineteenth century, which saw books like Edward Clarke's *Sex in Education; or, A Fair Chance for Girls* (1873), published at the height of women's demands for admission to universities in the United States. Women's desire to develop their intellect, Clarke argued, was the highest form of egoism, threatening to undermine the health of the race and to cause women's ovaries to shrivel.[12]

Given its history, should sexual differences be studied at all? In 1995 a group of women urged the University of Pennsylvania neuroscientist Raquel Gur to stop publicizing her studies on men's and women's brains for fear that the very notion that they are different might set women back twenty years.[13] Though scores of studies have been undertaken to show that women fail to measure up to men, it is astonishing how little we know about female bodies when it comes to keeping women healthy.

Historically, then, medical models of sexual difference have operated in several ways. "Sexual science" has typically used medical evidence to argue for women's social inequality using a paradigm of radical physical and intellectual difference. In medicine more generally, where health is at stake, research has vacillated between emphasizing sameness and difference. This legacy has led current researchers to assume either that women's and men's maladies are similar, when in fact they are not; or that men's and women's illnesses are different, when in fact they are similar. The paradigm of sameness has caused certain aspects of women's health to be understudied, for instance the interplay between estrogen therapy and cardiovascular diseases. The paradigm of radical difference has been prominent in diagnosis, where women's complaints are often dismissed as psychosomatic. (Higher proportions of women than of men are assigned diagnoses of "nonspecific symptoms and signs" in both health service records and death certificates.)[14]

Whether male and female bodies have been construed as paradigmatically similar or different, the male body has been taken as the primary object of research. Women's bodies have been considered a deviation from the male norm, and studies have focused on their reproductive uniqueness. The results of medical research conducted among men are then applied to women, even though outcomes for women in disease, diagnosis, prevention, and treatment in the nonreproductive sphere have not been adequately studied. Only recently have physicians become aware of how injurious to women's health neglecting research on women can be. A model of critical attention to sexual difference in relation to medical care now stands at the center of reforms in women's health research.

Correcting the Biomedical Model

The late 1980s saw a great awakening of mainstream medicine to women's health concerns. Feminist researchers criticized several large and influential studies that omitted women as both objects and subjects of medical research—most notably the 1982 Physicians' Health Study of Aspirin and Cardiovascular Disease, performed on 22,071 male physicians and 0 women; the Multiple Risk Factor Intervention Trial (now commonly known as MR. FIT), looking at the correlation between blood pressure, smoking, cholesterol, and coronary heart disease in 12,866 men and 0 women; and the Health Professionals Follow-Up Study of heart disease and coffee consumption in 45,589 men and again 0 women. There are many other examples. The National Institute on Aging's Baltimore Longitudinal Study of Aging, begun in 1958 and now considered the definitive report on "normal human aging," includes virtually no data on women, despite the fact that women constitute two-thirds of the population over age sixty-five. Most surprising of all, the first study of the role of estrogen in preventing heart disease was conducted solely on men (because the hormone was considered a possible treatment).[15]

Women's health concerns have not been entirely ignored. The Nurses' Health Study of the late 1980s followed 87,000 registered nurses for six years to study the correlation between taking aspirin and the risk of heart attack. Unlike the Physicians' Health Study, the original Nurses' Health Study was an observational investigation, not a more costly randomized clinical trial. Like the study of physicians, the study of nurses looked at predominantly white, health-conscious populations.[16]

The results of studies of men, the ensuing diagnoses, preventive measures, and treatments have commonly been extrapolated to women. It would be highly unusual for the results of a study of women to be assumed to apply to men.

Women have also been excluded from drug trials, even though women consume roughly 80 percent of pharmaceuticals in the United States. Until the spring of 1988 clinical trials of new drugs by the Food and Drug Administration (FDA) were routinely conducted exclusively on men. The results of these drug tests were then generalized to women, who were (and still are) typically prescribed dosages devised for men's average weights and metabolisms. Although little is known about the effects of aspirin on heart disease in women, women of the appropriate age have been encouraged to take an aspirin each day. Other widely used drugs, such as Valium, were never tested on women, although 2 million women per year take Valium. A 1992 study by the General Accounting Office found that only half the drugs surveyed had been analyzed for sex-related differences.[17] It is now known that acetaminophen, an ingredient in many pain relievers, is eliminated in women at about 60 percent the rate for men. Giving drugs to women at dosages designed for men puts women at risk for overdosing.

Investigators have defended their choice of men as research subjects on the grounds that men are cheaper and easier to study. Women's normal hormonal cycles are viewed as methodological problems complicating analysis and making it more costly; researchers have also feared that including women of childbearing age in clinical trials might endanger potential fetuses. (The FDA guidelines restricting research on women were implemented in 1977 in reaction to the birth defects from thalidomide and diethylstilbestrol—DES—taken during pregnancy and were lifted only in 1993.) These protections, however, portray women as "walking wombs," unable or unwilling to control their fertility, and ignore post- menopausal women. These measures also overlook the needs of many pregnant women, three-quarters of whom require drug therapy and currently use prescription or over-the-counter drugs for chronic conditions such as diabetes or depression.[18]

A 1981 review of studies of women's health found twice as much research on women in childbearing and childrearing roles than on other health problems. Despite this focus on reproductive health, not one of the more than fifteen institutes and centers that make up the NIH is devoted to gynecology and obstetrics. In the late 1980s the NIH, with only three obstetrician-gynecologists on its permanent staff, employed more veterinarians than gynecologists.[19] Ob-gyn has been part of the National Institute of Child Health and Human Development, where the focus is on the health of infants and children, not that of the women bearing them.

The net effect of gender bias in medical research and education is that women suffer unnecessarily and die. Adverse reactions to drugs occur twice as often in women as in men. Some clot-dissolving drugs used to treat heart attacks, for example, while beneficial to many men, cause bleeding problems in many women. Standard drugs for high blood pressure tend to lower men's mortality from heart attack but have been shown to increase deaths among women. Evidence is also emerging that the effect of antidepressants varies over the course of the menstrual cycle and thus that a constant dosage may be too high at some points in a woman's cycle and too low at others. Not only are drugs developed for men potentially dangerous for women; drugs potentially beneficial to women may be eliminated in early testing because the test group does not include women. At the same time that women tend to be undertreated in many areas of medicine, they are at risk for overtreatment in the area of reproduction, such as unnecessary Caesarean sections and hysterectomies.[20]

Many people do not want to be subjects of medical research. A preference for male research subjects may have been fed by the tendency to rely on readily available populations that for a variety of social reasons were largely male: medical students, prison inmates, military personnel, and patients at Veterans' Administration hospitals. The history of experimentation on African Americans makes minorities wary of the medical establishment. The forty-year Tuskegee Syphilis Study, in which the U.S. Public Health Service deliberately denied treatment to black men who had syphilis, is infamous. In a less well known incident, Dr. J. Marion Sims, often celebrated as the father of modern U.S. gynecology, experimented on slave women in the 1840s. In the days before anesthesia, these women endured up to thirty operations each as Dr. Sims explored ways to repair vesico-vaginal fistulas, tears in the tissues between the bladder and the vagina usually resulting from childbirth.[21]

Beginning in the late 1980s and 1990s feminist reform in publicly funded biomedical research in the United States was pushed forward by the federal government. In 1986 the NIH initiated a requirement that grant applications include female subjects in medical testing and research; these guidelines were reissued in 1987 with emphasis on including minorities. Though the guidelines were generally ignored, the NIH issued a mandate in 1990 to include women in all research and founded the Office of Research on Women's Health (ORWH) to oversee the process. In 1993

Congress passed the NIH Revitalization Act, making the ORWH a permanent part of NIH and mandating the inclusion of women and minorities in medical research. Further, women's health concerns are being addressed by the fourteen-year $625 million Women's Health Initiative, the largest single study ever undertaken by NIH. In 1993 the Food and Drug Administration also revised its guidelines to allow more women of childbearing age to participate in early-phase drug trials, and in 1994 it established the Office of Women's Health to correct gender disparities in drugs and testing policies.[22]

Much of the impetus for the women's health movement came from liberal feminism's notion that women should get their fair share of research dollars both as researchers and research subjects. Attention was drawn to the failure to include women in publicly funded research: because women pay taxes for health research, they deserve to derive benefit from that research.[23] Simply taking women seriously as makers of knowledge and as subjects of research on topics other than reproduction (a baseline liberal approach) has had a tremendous impact in medicine. The changes have been simple, but their results have been dramatic: women's right to inclusion in basic medical research is now secured by federal law.

Beyond the liberal approach emphasizing equal attention to men and women, a reconceptionalization of sexual differences in the human body has been important to advances in women's health. When the Government Accounting Office studied NIH policies in 1989, there was still no uniform definition of research on women's health. Medical researchers (not those interested in sexual science) had long assumed that "women's health" referred to reproductive health—involving attention to childbirth, contraception, abortion, breast and uterine cancers, premenstrual syndrome, and other maladies distinctively female. Florence Haseltine, a powerhouse for reform at NIH, has identified the shift from reproductive to more general health issues for women—the notion that women's distinctive physiology can make the difference between life and death—as crucial for current reforms in women's health research.[24]

NIH now defines research on women's health as the study of diseases found only in women (such as breast cancer), or diseases with a higher prevalence in women or some subgroup of women (such as osteoporosis), or diseases that present differently in women (such as heart disease).[25] Working from this conceptual base, the 1991 Women's Health Initiative has focused on the causes, treatment, and prevention of the three leading killers of postmenopausal women: cardiovascular disease, cancer, and osteoporosis. The NIH Office of Research on Women's Health has also funded understudied areas, including women's occupational health, gender differences in auto-immune diseases, and women's urologic health.

Not everyone agrees that women's health requires special attention. Critics deny that it has been improper to leave women out of randomized trials such as the MR. FIT study. According to this view, men die at earlier ages from heart disease and are consequently an appropriate group for study. Other critics charge that the $625 million earmarked for female-specific disorders is too much (the Women's Health Initiative currently receives about 6 percent of NIH's $7 billion annual budget). They argue that 13 percent of the NIH annual budget is already devoted to health issues directly related to women, such as breast and ovarian cancer, gynecological and obstetrical care, and osteoporosis, while only 6.5 percent goes to diseases unique to men. They also point out that on a per-fatality basis more than four times as much research money is spent on breast cancer as on prostate cancer. Their trump card is that U.S. women's life expectancy (78.6 years) substantially outstrips men's (71.8 years), suggesting that women are well cared for (they tend not to discuss how many years of life are free of disability). No group of American men for whom statistics are collected outlives any group of women: African-American women (73.5 years) and Hispanic women (77.1 years) have longer life expectancies than white men (72.7 years). With two out of every three health-care dollars spent on them, the argument goes, women can hardly complain that their health needs are ignored.[26]

Critics on the other side object to the notion that feminism has made sufficient headway within medicine and charge that the Women's Health Initiative and the poorly funded NIH Office of Women's Health Research are merely efforts to defuse the explosive politics surrounding women's health. Still other critics point out that inequalities in biomedical research are not what haunts the majority of the world's women. In many Third World countries the problem is high maternal mortality. The World Health Organization's Safe Motherhood Initiative, launched in 1987, has received paltry funding, these critics charge, and steps to improve medical conditions for women the world over are long overdue.[27]

What is equal or fair in this instance? Is the solution to equalize spending on research on men's and women's health? One might argue that research using the male body as the norm serves men better even when fewer dollars are spent on male-specific disease. One might also argue that women's greater role in reproduction warrants more research on female reproductive health. But surely the point is to study both men and women of various classes, races, and backgrounds so as to maximize their long-term health and well-being.

The Community Model

Feminist reform within NIH, carried through by Florence Haseltine and many others, has been critical to improving health care for women. But other feminists, including Adele Clarke, Elizabeth Fee, Vanessa Gamble, and Nancy Krieger, suggest that it may not be enough to add women to studies already in progress or to take account of women's distinctive physiology. Study populations can be reconfigured, negative images of women can be altered, women's diseases can be given priority within existing medical research—without dramatically improving women's health. These critics contrast the dominant "biomedical model" to the "community," "social," or "ecosocial" model for women's health. They challenge clinical and biomedical models, which focus narrowly on disease management and biochemical processes in organ systems, cells, or genes. Sex and race, they claim, are more than biological variables. Depression in women, for instance, is often blamed on hormonal disturbances, when it may in fact be produced or intensified by discrimination, poverty, abusive spouses, or chronic ill health.[28]

These broader social models that ground health in the community do not ignore genetic or biological aspects of health—certainly the genetic components of Tay-Sachs disease, sickle-cell anemia, cystic fibrosis, and thalassemia require study. Nor do these models underplay the importance of personal lifestyle (attention to nutrition, exercise, relaxation, and restraint from smoking and substance abuse). But they also take into account how health and disease are produced by a person's daily life, access to medical care, economic standing, and relation to his or her community. They see health as embedded in communities, not simply in individual bodies.[29] The conundrum concerning women's longevity, for example, is partially solved by investigation of social factors. Longevity in women may not result from superior genes, health care, or healthy living, but from not being male. In industrialized countries young men die from dangerous jobs, war, firearm injuries, motor vehicle crashes, and use of illicit drugs and alcohol—hazards related not to biological frailties but to occupation and codes of masculinity. Older men die of heart disease, which may also be related to occupation.

Social factors also play a role in hypertension, which in the United States has been studied in predominantly white male populations, even though high blood pressure is more common among African-American women and men. At least one factor involved in hypertension is a person's exposure and response to discrimination. A black woman who protests unfair treatment, for example, is less likely to suffer higher blood pressure than one who quietly turns the other cheek. Ironically, the African Americans at highest risk for hypertension are working-class women and men who claim they do not suffer from racial discrimination.[30]

Much has been made of the need to depart from the "usual male model" and from the "usual white model" in medical research and health. Feminists are now wary of developing a "usual female model." Whereas the women's health movement of the 1970s sought to solidify sisterhood through the commonalities of women's childbirth experiences, many feminists now emphasize the differing health needs of different groups of women. African-American women, for example, are more at risk for stroke, heart attack, and hypertension than European-American women. While African-American women have lower rates of breast cancer than European-American women, they more often die from it. Hispanic women have rates of cervical cancer twice as high as non-Hispanic white women. Non-Hispanic white women have higher rates of osteoporosis than Hispanic or African-American women. Because osteoporosis is considered a white disease in the United States, African-American and Hispanic women may not be properly screened and educated about it.[31]

Looking again at hypertension, poor women are at greater risk than affluent women; within each income level, black women are more likely to be hypertensive than white women. The United States collects health statistics for five racial or ethnic categories: American Indians, Hispanics, blacks, Asian and Pacific Islanders, and whites. These unrefined categories ignore critical distinctions within ethnic groups. "Hispanic" women's risk for high blood pressure, for example, varies by national origin: it is lowest for Mexican women, higher for women from Central America, and highest for Puerto Rican and Cuban women. The same variance occurs among "Asian and Pacific Islanders": Japanese- and Chinese-American women generally have low blood pressure, while many Filipina women suffer from hypertension. Rates for hypertension also vary among Native American women: those who live in the northern plains have higher rates than those in the Southwest.[32]

While U.S. health statistics provide information on race, they rarely provide information on class, thus presenting the impression that the health profile of an African-American professor of law is similar to that of an African-American welfare mother. Matters of class often outweigh matters of gender, race, or ethnicity. Education, which correlates highly with socioeconomic status, has clear effects on health for both women and men. Death rates for women of all races aged 25–64 who did not finish high school are twice as high as those for women who did.[33]

Medical school training is also a crucial part of the formula for changing U.S. health care for women. The Office of Research on Women's Health has begun surveying medical school curricula, analyzing their approaches to women's health. New approaches to health may require new relationships between medical disciplines. In a profession driven by certifications, the newly founded American College of Women's Health is looking for board certification for a women's health specialty. Critics have worried that a special focus on women's health might ghettoize these issues and produce a group of low-paid female researchers and practitioners, while the rest of the medical profession would continue to practice "medicine as usual." This, of course, has been the danger of academic women's studies in general—that gender studies has become associated exclusively with women both as practitioners and as research subjects.[34]

Women's health centers, now part of many medical schools, are one result of efforts to restructure the profession. In the past many of women's health needs got lost in the gap between gynecology and other specialties. Incontinence, for example, was traditionally ignored by both urology and gynecology. A new specialty, urogynecology, developed in the 1990s, may offer women help in this area. Women's health centers are big moneymakers, and they repattern health care so that a healthy woman need not pay regular visits to two doctors: an internist for her "neutral" or non-reproductive parts and a gynecologist for her "womanly" or reproductive parts.

At present little scholarship in this area looks beyond North America or Europe; hence biomedicine may take the female condition in Western cultures as characteristic of the female condition in general. Biomedicine tends to ignore how culture, ecology, and economics can alter basic physiology. This becomes even more pronounced across cultures. Take the example of menopause. Cultural anthropologists, such as Susan Sperling and Yewoubdar Beyene, are demonstrating that women's physical experiences differ in response to cultural values, nutrition levels, marriage patterns, and so forth. In Western societies, the average age at menarche is thirteen and the average age at menopause is fifty-one. The women in these societies, who tend to marry late, have few children, and breastfeed briefly, typically experience thirty-five years of ovulatory cycles. Women in non-industrialized societies, by contrast, typically go through menarche at around age seventeen and menopause at forty-two. Long years of lactation (in an environment of moderate nutrition) inhibit ovulation, so that many women in these societies have a total of forty-eight menstrual cycles in a lifetime, or approximately four years of cycling.[35]

These differing physiological manifestations are accompanied by differing cultural meanings of menopause. The anthropologist Margaret Lock has found that in North America menopause is widely seen in terms of pathology and crisis—as a "deficiency disease" or "endocrinopathy"—while in Japan it goes relatively unmarked. Japanese women rarely suffer hot flashes; they more often complain of stiff shoulders, headaches, tiredness, and dizziness—symptoms generally treated without medicines. The rough Japanese term for menopause, *kōnenki,* does not refer to the cessation of menstruation but to a distinct period of life when the body lacks harmony in the autonomic nervous system; herbal regimes are sometimes recommended to help restore balance.[36] Doctors in Japan have not been concerned about hormone replacement therapy—partly because the incidence of osteoporosis among Japanese women is less than half that among white women in North America and mortality from heart disease is about one-quarter (even though Japanese women have one of the longest life expectancies in the world). Hormone replacement therapy is sometimes used as a preventive measure against stroke, a problem for elderly women and men in Japan.

Medical anthropologists caution against universalizing Western patterns of ovulation and menopause in theoretical models and clinical practices concerning hormone replacement therapy. The North American and European pattern of constant ovarian cycling may not be a norm of female physiology. Sperling and Beyene suggest that awareness of the differences in women's hormonal regimes around the world may lead to new ways of treating postmenopausal osteoporosis and cardiovascular risk.

What Brought Success?

A common assumption in research on gender in science is that the entry of women will change science. The problem with this assumption is that it is reductive and simplistic. Medical research in the United States has undergone a sea change with respect to women: health research has become more responsive to women's needs, and women's health centers represent a new approach to women's health care. What, in addition to the increasing numbers of women in medical professions, contributed to this success in the medical sciences? Can the medical sciences be used as a model for reform in other sciences?

The women's health movement emerged in the 1960s and 1970s. Local and national groups—including the Boston Women's Health Book Collective, the National Women's Health Network, and later the National Black Women's Health Project and consumer lobbies for the treatment and prevention of breast cancer—began drawing attention to how the U.S. health care system failed women. Activists in these groups questioned male control of the health professions, encouraged women to enroll in medical schools, challenged sexism in traditional medical curricula, and fought to license midwives and to improve women's knowledge of their bodies.[37]

This movement benefited from broader social change that put into place a host of legislation, such as the Equal Employment Opportunity Act of 1972 and the Equal Opportunity in Science and Engineering Act of 1980, which specifically directed NIH and the National Science Foundation to increase the participation of underrepresented groups in medicine, science, and engineering. The women's health movement was shored up by affirmative action policies that fostered equal opportunity for women and minorities within universities and industries doing business with the federal government.[38]

The women's health movement also benefited from the development of academic women's studies. Sociologists, anthropologists, and historians challenged the notion that, especially for women, "biology is destiny"; they undercut images of women as mentally and physically frail; they analyzed historical conceptions of women and their place in society.[39] Historians of medicine documented that the 165-pound white male body served as the "gold standard" for medical research and treatment, showing, for example, that standard medical textbooks discussed women primarily in sections on reproduction, while discussion of nonreproductive parts—kidneys, respiratory system, stomach, and so forth—focused on men.[40] A solid body of research and well-honed tools of analysis were in place as reform moved forward.

Reform of NIH policies concerning research on women's health also depended on a significant number of people dedicated to women's issues who were well placed within the medical profession and within NIH itself. Florence Haseltine, director of the NIH Center for Population Research, founded the Society for the Advancement of Women's Health Research in 1990; William Harlan, then director of epidemiology and clinical applications at NIH, supported (and coined the name of) the Women's Health Initiative. Ruth Kirchstein, now deputy director of NIH, served as the first director of the Office of Women's Health Research. The Women's Health Initiative, the largest research program ever undertaken by NIH, took shape when Bernadine Healy, a Bush appointee, became NIH's first woman director. By 1986 NIH had established an Advisory Committee on Women's Health Issues and recommended increasing women's participation in federally funded biomedical research.

The most striking aspect of the reforms, and what sets medicine apart from other areas of science in which feminists have sought change, is the intervention of Congress. The powerful Congressional Caucus on Women's Issues, with its office, six-member staff, and $250,000 annual budget was staffed by articulate and powerful women, such as the Democrats Patricia Schroeder and Barbara Mikulski and the Republicans Constance Morella and Olympia Snowe, who have served as ardent advocates of health reform. They were joined by active lobbyists, especially for breast cancer research, who brought national attention to that emotional issue. In 1989 the Women's Caucus introduced the Women's Health Equity Act (WHEA) bill, modeled on the Economic Equity Act. The WHEA called for establishing a permanent óffice of women and health under the assistant secretary of health, charged with overseeing the inclusion of women in research studies and providing money for research into several diseases specific to women. While several of the proposals in WHEA became law (the Breast and Cervical Cancer Mortality Prevention Act, which provided mammograms and Pap smears to low-income women and Medicare coverage for screening mammography), WHEA as a whole did not.[41]

In June 1990 the Caucus on Women's Issues, joined by Henry Waxman, chair of the powerful House Energy and Commerce Subcommittee on Health and Environment, called for the General Accounting Office (GAO) to investigate whether NIH had implemented its own 1986 policies concerning women. The GAO's extremely critical report sparked NIH to found the Office of Research on Women's Health and the Women's Health Initiative.[42] These efforts were aided by shrewd policymakers who used the NIH reauthorization bill (which happened to be up for renewal) to move their legislation forward. The reform was also aided by a climate of opinion that had elected a Democratic president concerned about health and health reform; earlier legislation of this nature had been vetoed by President Bush.

The end of the Cold War also contributed, as a portion of the mammoth defense budget was turned toward women's health research. By 1995 some $400 million had been appropriated for the army's research on the genetics of breast cancer and for research on ovarian cancer and osteoporosis. The Department of Defense also provided funds for the Defense Women's Health Research Program, which analyzes the health care needs of military women.

Finally, the women's health movement also benefited from the emergence of a new class of professional women, ready to talk openly about and demand care for disorders, such as incontinence, long hidden behind the cloak of female modesty. Well-educated women demanded new relationships with doctors, formed networks for exchanging information, and took charge of their health needs.

These and many other changes in attitudes toward women contributed to the beginning of reform in research on women's health. To see the entry of women into the profession as the single decisive factor is to oversimplify and depoliticize a very political and complex historical process. Many have observed that concern about gender bias in research arose only as women began to occupy the senior levels of scientific institutions and Congress. Women in the profession, of course, are important. The NIH Office of Research on Women's Health has rightly set recruiting and retaining women in biomedical careers as an important priority along with strengthening research on diseases that affect women and seeing that women are adequately represented in biomedical and behavioral studies. There is a need to continue to promote women in medical careers. In 1993 only 18 percent of the tenured scientists at NIH were women. Nationwide, in 1994 women constituted 1 percent of medical school deans, 4 percent of heads of medical school departments, and 22 percent of the medical school faculty members.[43]

Getting women into the profession, however, is only one aspect of reform. Changes in medical practices and research required a broad-based women's movement, fundamental changes in attitudes toward women and their place in society and the professions, the institutionalization of academic research on gender, strong lobbies on explosive issues (such as breast cancer), a reasonably strong economy (and the end of the Cold

War), a climate of opinion that elected a Democratic president, and finally an act of Congress. The same forces that brought women into the profession also allowed for change in research questions relating to women. It was not just women but feminists—both men and women—inside and outside medicine who created the conditions for success of the reforms in medical research.

7

Primatology, Archaeology, and Human Origins

Lucy, a 3.2-million-year-old fossil hominid from Ethiopia, was declared female when "she" was unearthed in 1974. How and why was Lucy judged to be female? Why, the paleontologist Lori Hager asks, was the discovery named "Lucy" ("in the Sky with Diamonds") and not "Sergeant Pepper" (from the equally well known Beatles song)? The sex of an individual is commonly determined by genitalia (soft tissue that does not fossilize) and DNA (which is rarely isolated in fossils without contamination). Where these materials are not available, sex is determined by pelvic morphology, body size, and, in nonhumans, canine teeth. Upon finding "Lucy," Donald Johanson and his team sexed the fossil female: you can tell a female, they wrote, because "the pelvic opening in hominids has to be proportionately larger in females than in males to allow for the birth of larger-brained infants." Lucy and her kind, however, did not have large brains; nor, on closer inspection, did Lucy have the larger pelvis to permit the birth of larger-headed infants. Distinctively large hominid brains did not develop until roughly 2 million years ago. Despite this seeming contradiction, Johanson gave no other evidence for considering his prized fossil "undoubtedly female."[1]

This story highlights some of the biases that have driven studies of human origins. Lucy is often considered female simply because of her size; she stood three feet seven inches tall and is thought to have weighed no more than sixty pounds. But perhaps "she" was, in life, a male of a diminutive species rather than a small female of a species with large differences in size between the largest (presumably male) and the smallest (presumably female) individuals. Paleoarchaeologists recognize that the small and large fragmentary fossils found in East Africa can represent either female and male members of one highly sexually dimorphic species or individuals belonging to two different species, one large and the other small. Despite this uncertainty, little Lucy is usually considered female. This is the view promoted by the permanent exhibit entitled "Human Biology and Evolution" that opened to much fanfare at the American Museum of Natural History in New York in 1993. The *Australopithecus afarensis* diorama breaths life into the bones, re-creating a robust male who towers over a smaller consort, his arm positioned to protect and reassure her (Figure 5). Though 3.5-million-year-old footprints preserved in lava near Laetoli, Tanzania, clearly show two individuals walking together in stride, they may not be the male and female nucleus of a modern family—they may be a parent comforting his or her adolescent offspring or just two friends fleeing the volcano together.[2]

Such assumptions about humankind's distant ancestors have been challenged by feminists in the fields of archaeology, paleoanthropology, and evolutionary biology. What new insights have feminists brought to questions concerning human origins and evolution? *Therefore, feminist are ousveloniyosome of these assumptions.*

Primatology (?)

Women have done well professionally in primatology, in which they now receive close to 80 percent of all Ph.D.'s. This is astonishing given that no Ph.D.'s at all were awarded to women in this field in the 1960s. Women already garnered over 50 percent of the Ph.D.'s in primatology in the 1970s, and this rose to 60 percent in the 1980s and 78 percent today.[3]

Primatology is widely celebrated as a feminist science, or at least as a field in which women have refitted fundamental paradigms. The extraordinary focus on *women* as agents of change has been cultivated not least by women primatologists themselves, beginning in 1984 with Meredith Small's book *Female Primates: Studies by Women Primatologists*. While few female primatologists call themselves feminists, most do not deny that much of their scholarship has been motivated by feminist concerns. And they have taken up the question of the impact of the women's movement on their discipline, a development that is also taking place in anthropology, archaeology, and evolutionary biology but in few other branches of science.[4] Is primatology a feminist science?

Nearly everyone involved with this issue agrees that after World War II primatology was rife with stereotypical attitudes toward males and

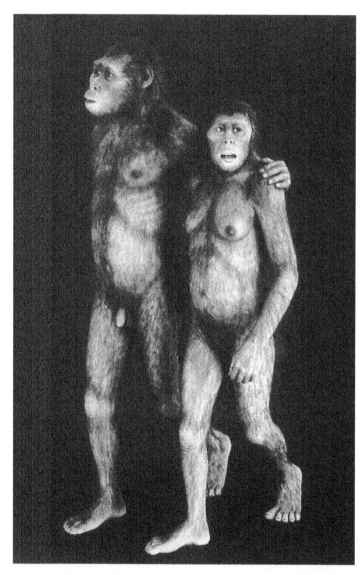

Figure 5. Reconstructions of the early humans presumed to have made the Laetoli footprints, as shown at the American Museum of Natural History in New York—fact or fantasy? Courtesy Department of Library Services, American Museum of Natural History.

females. Primatologists tended to divide primates into three groups for study: dominant males, females and young, and peripheral males. These divisions reinforced the notion that primate society was driven by competition among dominant males who controlled territorial boundaries and maintained order among lesser males. Females (often studied with the young as a single reproductive unit) were described as dedicated mothers to small infants and sexually available to males in order of the males' dominance rank, but otherwise of little social significance (Figure 6).[5] Primatologists tended to view females as noncompetitive, docile creatures who traded sex and reproduction for protection and food.[*]

The feminist remaking of primatology, like that of medicine, hinged on analyses of the ways choices of study subjects can influence the results of science. In this instance the choice of subjects went beyond looking for a representative mix of males and females. The primatologist Linda Fedigan has discussed the 1950s myth of the "killer ape," the pervasive image of primates engaged in a Hobbesian war of all against all, a vision seething with dark implications for human nature. This image of aggressive primates was drawn almost exclusively from studies of savanna baboons; Fedigan has called this process the "baboonization" of primate life. Male baboons are typically portrayed as given to bullying of females and violent infighting with other males (Figure 7). From the 1950s to the 1970s baboons were the most widely studied monkeys, despite widespread knowledge that other species could provide more sanguine visions of ancestral humans.[6] *was it because baboons were a much bigger animal?*

Why, despite the alternatives, should baboons and other aggressive populations have dominated postwar primate studies? For one thing, baboons live on the ground, making them accessible to humans (90 percent of primate species are arboreal).[7] Second, they inhabit the African savanna, considered the birthplace of "early man," and they were thought to share certain selective pressures with protohominids. Equally important, the image of primate society as aggressive, competitive, and male dominated played well to a public embroiled in the Cold War. Baboons provided a ready explanation for human warfare, violence, and male aggression. In this instance the choice of subject matter introduced a potent antifeminist element into primatology, highlighting and reinforcing notions about male dominance.

In primatology, as in medicine, the majority of feminist changes to date have come from reevaluations of females. Only in the 1960s did primatologists begin looking seriously at what females do. Feminists first over-

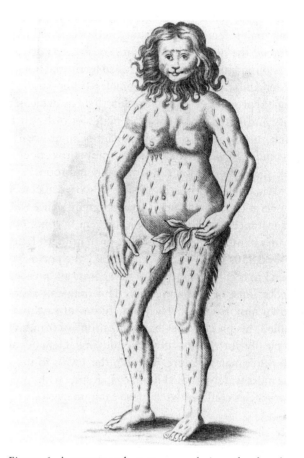

Figure 6. A seventeenth-century rendering of a female "orang-outang," her privates covered by a fig leaf to protect her very great modesty. Source: Edward Tyson, *Orang-Outang, sive Homo Sylvestris; or, The Anatomy of a Pygmie Compared with That of a Monkey, an Ape, and a Man* (London, 1699). By permission of the National Library of Medicine.

turned the conventional stereotype of the passive, dependent female. Jane Lancaster's article "In Praise of the Achieving Female Monkey" (1973) began with a notion, radical for its time, that females "can too"—that anything males can do, females can do.[8]

In many instances reevaluation of gender stereotypes went beyond the liberal paradigm of "sameness" to a new examination of sexual difference.

Figure 7. Since the seventeenth century the baboon has symbolized male virility and aggression. Source: Thomas Bartholin's *Acta Medica & Philosophica Hafniensia* (Copenhagen, 1673), vol. 1. By permission of the National Library of Medicine.

Seeing the primate world "from the female monkey's point of view" (as the Berkeley zoologist Thelma Rowell has put it) called into question many fundamental assumptions about the primate world. Primatologists questioned stereotypes of male aggression, dominance and alliance, and female compliance. They studied the significance of female bonding *female bonding by*

1.) to dates

through matrilineal networks, analyzed female sexual assertiveness, female social strategies, female cognitive skills, and female competition for reproductive success. Rowell found, for example, that older female baboons determined the route of daily foraging; Shirley Strum found that male investment in "special relationships" with females had greater reproductive payoff than did a male's rank in a dominance hierarchy. Today, in a turnabout from the 1960s, conventional wisdom concerning baboons recognizes that females provide social stability while males move from group to group. While this phase of critique did not produce alternative explanatory paradigms for the discipline, it cast suspicion on key conceptions about aggression, reproductive access, and dominance.[9]

The impulse to look at primate populations from a "female point of view" found a curious bedfellow in 1980s-style sociobiology. Sociobiology, widely acknowledged as a central paradigm in primatology from the mid-1970s to the mid-1980s, functioned early on as an antidote to feminism: if fundamentals, such as the sexual division of labor, are hardwired in the species, efforts to counteract them are foolhardy. According to the Harvard University professor E. O. Wilson, "men forage for game or its symbolic equivalent in the form of barter and money," while women seek the man with the best genes and subsequently tend his young. Donna Haraway has pointed out that Sarah Hrdy, a self-identified feminist, was among the first to apply what became known as sociobiological theory to primates. In primate studies at least, sociobiology, a theory widely excoriated by feminists, began female-centered and written by a woman.[10]

Many feminist innovations have occurred within theories of sexual selection, which is seen, after natural selection, as a prime motor of biological evolution, and which Sarah Hrdy has dubbed the "crown jewel" of sociobiology. Darwin traced secondary sexual differences to the cosmic drama of sexual selection. Certain characteristics—Darwin mentioned brilliant plumage, heavy horns, courage, pugnacity, perseverance, strength and size of body, weapons of all kinds, musical organs, bright colors, ornamental appendages—are selected and perpetuated because they lend an individual of one sex, "generally the male," an advantage in his struggle for access to the female and allow him to leave a larger number of offspring "to inherit his superiority." Darwin and others long assumed that sexual selection does not act as strongly on females as on males; consequently they emphasized male-male competition for females and female mate choice as the mechanisms of selection. Males are the wooers; females, "though comparatively passive, generally exert some choice" in accepting one of the victorious males (Darwin exempted humans from the latter practice because, as was clear to his fellow Victorians, human males proposed marriage). The notion that males are competitive and females are coy has been so persuasive that for more than twenty years a group of ornithologists searched for the "alpha males" within a population of pinyon jays, even to the extent of setting up limited feeding stations to inflame competition. As it turned out, however, the birds that lock talons and bills in deadly combat in this group are females.[11]

Ignoring ardent female-female competition over nutrients or nest sites (as in the example of the pinyon jays) and choosiness among males (selecting as mates females with abundant parenting skills, high dominance status, good health, or good foraging ability) can skew notions of how evolution works. Another way to distort notions of sexual selection is to ignore interactions between males and females that go beyond the strict interpretation of sex as for reproduction only. Take for example Sarah Hrdy's female monkeys (savanna baboons, chimpanzees, South American tamarins, and other troop-dwelling species) "who forgot to be coy." Contradicting stereotypes of passivity among their kind, these females promiscuously pursued males, seeking copulation beyond what was necessary for fertilization. There are many reasons why females actively pursue "extramarital affairs" (in sexual selection parlance "extrapair copulations"); Hrdy focuses on the need for females to win from males parental care for their offspring.[12]

Feminist sociobiologists (for some an oxymoron) have been harshly criticized by other feminists. Within primate studies, sociobiologists have been accused of producing the "corporate primate": female baboons with briefcases, strategically competitive and aggressive. Female apes and monkeys have been observed forming stable dominance hierarchies and alliances with males other than their mates, displaying aggression, exercising sexual choice, and competing for resources, mates, and territory much like males. Females emerged as newly enfranchised citizens in nonhuman primate society as feminists began reevaluating female primates in terms of traditional male behaviors. In the feminist-sociobiological narratives, both males and females are aggressive, competitive, and struggling for genetic advantages.[13]

Feminist sociobiologists have retold the story of evolution, remaking females into active participants, but some critics claim they have generally done so without changing the underlying theory. Fundamental concepts of sociobiology, such as reproductive success, however, may not allow for explanations of social behavior in biological terms; the current tendency in sociobiological theory to focus on the survival and replication of genes rather than on whole organisms reduces the notion of "social behavior" to reproduction and makes it an abstract category that is assumed to be homologous across classes of animals from insects to humans. The evolutionary biologist Patricia Gowaty has countered that theories of natural and sexual selection are not deterministic in ways that some feminists claim. She suggests that analyses of sexist behavior in terms of Darwinian natural selection might aid in overcoming sexism, and she emphasizes that while there may be a few essential differences between the sexes (menstruation, childbearing, lactation) there are strong differences in the selective pressures facing males and females and that these are social as well as biological. Linda Fedigan, whose empirical work has provided a foundation for recent advances in primatology, has taken a different tack, suggesting that mainstream primatology is a feminist science.[14]

Feminists have long sought the holy grail of feminist science; it is worth looking at Fedigan's intriguing argument in some detail. Drawing from a variety of feminist debates, Fedigan identifies six features of feminist science that she also finds characteristic of contemporary mainstream primatology. The first is "reflexivity": a sensitivity to context and cultural bias in scientific work. She contends that warnings issued in graduate school against anthropomorphism (assimilating apes to human behaviors, values, and motivations) and ethnocentrism (assuming that one's own culture is superior to others) cultivate reflexivity in primatologists. A second common feature is a critical attention to "the female point of view." A third is a respect for nature and an ethic of cooperation with nature. Because many nonhuman primates are endangered, Fedigan says, a great many primatologists are environmentalists, concerned about preserving primates and their habitats.

A fourth feature Fedigan identifies is the move away from reductionism. Primatology, she maintains, has moved away from viewing primates as responding primarily to genetic or hormonal directives and toward appreciating them as sentient and intelligent beings living within a complex set of social relations and traditions. Fedigan also argues that both primatology and feminist science seek to promote humanitarian values rather than national interests, though she warns that this is an impression and not the result of empirical study. The final feature common to feminist science and primatology is a scientific community that is diverse, accessible, and egalitarian; Fedigan notes that North American and European primatologists have begun to open their discipline to people of different nationalities, especially people from the countries their primate populations inhabit.

Only two of Fedigan's six elements have strictly to do with feminism: discussion of the politics of participation (that is, who is included and excluded from the community of scientists) and attention to females as research subjects (a critical appraisal of their role in primate societies and to gender in research paradigms). It is true that feminists are rarely advocates only for women. Many feminists see that incorporating women and their concerns into science cannot be done in isolation from other humanitarian and environmental concerns.[15] Yet compressing all these concerns into feminism can lead to untenable positions, as when ecofeminists assert that women have a special relationship with nature or that the oppression of women is somehow tied to the domination of nature.[16] Feminism has suffered from attempts to claim too much—all that is good and true—in its name. While no one would contest the assertion that a feminist might also be an environmentalist or a humanitarian, or might embrace a salutary self-consciousness about his or her assumptions and methods, these elements go beyond feminism as such.

Fedigan sees herself as a reporter or translator, working at the intersection between gender studies of science and mainstream primatology. In this role she reports, accurately I think, her findings on feminist science.

Fedigan's paper and its title question, "is primatology a feminist science," were prompted, I suspect, by Donna Haraway's massive and complex *Primate Visions* (1989). In Haraway's history of postwar primatology she picked up on the notion—put forth by primatologists themselves—that women were making a difference. While Haraway documented how some influential women (some self-identified as feminists, some not) challenged fundamental paradigms, she also emphasized that science is constituted by a multitude of factors, ranging from attitudes toward gender roles and domestic issues surrounding race and class to economic relations between the First World countries and the Third World countries where most nonhuman primates live.

Although many primatologists reacted negatively to Haraway's book—perhaps because her deconstructionist analysis challenged the authority of scientists, or because she was an outsider, or simply because the book was full of a highly literary poetics more likely to appeal to humanists—a number of primatologists are beginning to undertake, in modified form, analysis akin to Haraway's. Shirley Strum and Linda Fedigan have charted gender analysis within four "stages" (or one might say eras) of modern primatology: natural historical (1950–1965); structural-functional (1965–1975); sociobiological (1975–1985); and socioecological (1985–present).[17] While Strum and Fedigan distinguish among stages of primatology, they do not distinguish among stages or types of feminisms that they see as informing primatology.

Although Strum and Fedigan's style and approach differ from Haraway's, they come to some similar conclusions. They do not see women or feminism as the single factor (and sometimes not as the primary factor) leading to female-friendly change in primatology. They reject any simple-minded attempt to find a one-to-one correlation between women's entry into the field and feminist impulses. They also reject the idea that the systematic study of feminism's effects on science is a political question, peripheral to science itself. Many practicing scientists assume (wrongly) that feminism is something imposed upon science from the outside; Strum and Fedigan show that working primatologists have not merely studied how feminism has transformed primatology, many of them are feminists who have helped create that transformation.[18] Part of Strum and Fedigan's argument is that feminism has been central to the development of primatology and that feminist contributions should be studied as part of the history of the discipline.

While the extent of feminism's impact on primatology may remain in dispute, it seems clear that females are no longer considered secondary to the process of evolution. Since the work of Jeanne Altmann, Linda Fedigan, and Sarah Hrdy, females have been recognized as having their unique place in primate societies, and the ecology of female primate relationships has become a vigorous area of research.[19]

Human Evolution

The controversies in primatology from the mid-1970s to the mid-1980s were paralleled by similar developments in paleoanthropology, where the "woman the gatherer" hypothesis challenged the entrenched "man the hunter" thesis. For Sherwood Washburn and his peers in the 1960s, the hunting hypothesis explained how quadrupedal apes evolved into bipedal articulate toolmakers with significantly larger brains. The man-the-hunter hypothesis coexisted peacefully with the dominant baboon model, the former a worthy descendant of the latter. Evolutionary theory was sharply focused on males, giving the impression that men "evolved by hunting while sedentary women tagged along gathering and giving birth." Bringing home the wildebeest became the "master behavior" of the human species. Men actively and aggressively drove human evolution forward; only what Darwin had called the "equal transmission of characters" allowed traits selected for in males to be transmitted to females. Prehistoric women were rendered as invisible handmaidens to men.[20]

In the 1970s the anthropologists Sally (Linton) Slocum, Nancy Tanner, and Adrienne Zihlman developed the influential "woman the gatherer" theory of human evolution, arguing that women's foraging after wild plants, not men's hunting, provided the primary source of subsistence for the earliest humans. The gathering hypothesis saw women as active, not passive, participants in human evolution: as contributing to subsistence; as contributing to technological innovations associated with collecting, carrying, and sharing food; as contributing to social life through their centrality in reproduction and as conveyers of tradition from one generation to the next. Further, this new hypothesis undermined the notion that early human societies were characterized by strictly observed monogamy and rigid sexual divisions of labor with females subordinate to males. In contrast, the gathering hypothesis suggested that females, too, actively chose mates and that hominid societies were built around flexible sex roles with activities probably varying by age and reproductive stage of males and females rather than strictly by sex.[21]

Adrienne Zihlman, one architect of these developments, has emphasized that questions concerning women's role in evolution emerged in the context of the 1970s women's movement with its insistence on making women "visible." She has also carefully pointed out the role played by certain men, such as Richard Lee (not a feminist, in this case), in providing crucial new data about women's significant contributions to human well-being in hunter-gatherer societies. And she has argued that it is a mistake to suggest that the woman-the-gatherer hypothesis was developed in the context of feminist theory. There was no such theory available in the early 1970s. The feminist social climate provided, she says, "the basis for asking questions, but it did NOT provide data." What she, Linton, and Tanner did was to provide a new hypothesis for organizing newly emerging data—data that cast doubt on key aspects of the man-the-hunter synthesis. These data included findings like Lee's from the !Kung that "the women provide two or three times as much food by weight as the men," findings from new chimpanzee studies along with the new data revealing the genetic closeness of humans to chimps, and findings from the fossil record.[22]

The woman-the-gatherer thesis also rested on Sally Slocum's challenge to the definition of a "tool." Slocum rejected the idea that tools, defined as projectiles, knives, and axes, represented the earliest signs of human civilization. Richard Lee, fresh from his study of the !Kung and his observations of women's active role as gatherers and hunters of small prey, stressed that wooden digging sticks and skin containers used for gathering would not be preserved in the archaeological record. Slocum revised the category "first tools" into the broader notion of "cultural inventions" to draw attention to digging sticks, baskets (used for gathering), and slings (for carrying babies)—artifacts thought to have emerged from the female side of life. This reevaluation was bolstered by the absence of hunting tools among the earliest stone tools (around 2 million years old) found in the Olduvai Gorge and Koobi Fora. Hunting tools appear in the fossil record from half a million years ago.[23]

The woman-the-gatherer thesis has been criticized by some feminists for not going far enough. In spite of its revolutionary new perspective, Jane Balme and Wendy Beck have pointed out, "the rationale for the division of labor remained unchanged, men hunt and women gather because they are constrained by their reproductive roles." The main difference was that it was not hunting but gathering activities—particularly the practice of gathering plant food for delayed consumption—that generated human traits and inventions. But the woman-the-gatherer hypothesis did not challenge the duality of man the hunter versus woman the gatherer, so deeply ingrained in Western ways of thinking. To change the story in fundamental ways, the Berkeley anthropologist Margaret Conkey has called for a deeper critique of underlying assumptions. What does it mean, she asks, to attribute "the sexual division of labor" to apes or early hominids? The debate about man the hunter versus woman the gatherer is, she remarks, really about the origins of two Western social institutions: the nuclear family and a gender-based division of labor. To seek their origins is to accept these institutions as natural and legitimate, rather than to see them as products of particular histories.[24]

Conkey is correct to question origin stories: those who study prehistory should be critically aware of why they seek the "origins" of certain cultural arrangements (such as marriage, the family, and sexual differences) and not others. As we shall see in the story of archaeology, however, shifting the perspective to plants and gathering has led to other important innovations.

In contrast to the situation in primatology, the feminist attempts in the 1970s to retool accounts of human evolution were not to flourish. Adrienne Zihlman has thoughtfully discussed the fact that since the 1980s new perspectives on women's role in evolution have not been built into a new theoretical synthesis but have been largely dismissed. Owen Lovejoy's concept of "man the provisioner" even co-opted gathering as a male activity. Linking bipedalism with increased fertility and survival, Lovejoy hypothesized that human success depended upon an increase in hominid population size, achieved by a decrease in the interval between births. This was accomplished by a curtailment of female mobility. Lovejoy's paradigm reinstated rigid sexual divisions of labor: women once again were seen as immobile and continually breeding, dependent now on "scavenging," rather than "hunting," men. Zihlman notes that the work of Lovejoy and others both demeaned the contributions of women scientists and undermined the improving status of women as active participants in the drama of human evolution.[25] Zihlman's evaluation raises the question of why feminism has not enjoyed the same success in evolution studies as in related fields such as primatology, anthropology, and history.

Archaeology

A short item entitled "The Female Anthropologist's Guide to Academic Pitfalls," published in the *Anthropology Newsletter* in 1971, gives women the following advice:

Pick a field or branch wherein you can function independently. Areas demanding "team type" research are out, unless, of course, you are married to the field director, an ideal situation, and one devoutly to be recommended. Such beleaguered fields include archaeology, serology, and genetic and medical anthropology, all of which demand the cooperation and/or participation of large numbers of colleagues or professionals in related fields. Men seldom care to include a woman (except as cook or technician) on their expeditions . . . Instead, choose any branch of ethnography, linguistics, musicology, primate behavior, anatomy, nutrition, computer science, etc., in which you can undertake research alone or with a compatible colleague.[26]

Intending to be ironic, this notice has instead proven prophetic. Women have excelled in primatology, but have been less prominent in paleontological fieldwork, one celebrated exception being the legendary Mary Leakey, who was, sure enough, married to the field director.

Feminism came curiously late to archaeology, given archaeology's close affinities to anthropology, ethnography, and history, in which gender studies have been influential since the 1970s. Perhaps it was insurmountable disciplinary boundaries that insulated archaeology from gender analysis for so long. Or, as Conkey has suggested, perhaps the strong positivist methods governing the field discouraged the self-reflection characteristic of gender analysis. Since the 1990s, however, feminist archaeologists have come on strong, and have published a comprehensive book on equity issues in the profession and several books analyzing the content of the science.[27] Feminist archaeologists have insisted, more than other scientists, on the relationship between the gendered structure of their discipline and the knowledge produced. Here I want to highlight their findings on how status hierarchies in the discipline have subordinated women as both subjects and objects of archaeological investigation.

Margaret Conkey and Sarah Williams open their analysis of the "political economy of gender in archaeology" (1991) by throwing into question a traditional object of archaeological knowledge: origin stories. The search for origins—of hominids, the state, agriculture, trade, fire, gender roles, status, toolmaking, hunting, language, and so forth—defines the "big," prestigious questions in archaeology. The primacy of research on origins, Conkey and Williams claim, allows its practitioners to structure the discipline, influence career success, and make political statements about human nature and human society in presenting the results of their research.[28]

As is well known, research on origins has traditionally left little place for women or gender analysis in the evolution of humankind. Fossil women, even when imagined to have contributed to cultural innovations, are seldom portrayed as having made major strides in human evolution. What is new in Conkey and Williams's discussion is their recognition of the role played by gender inequalities in intellectual authority in determining what counts as archaeological evidence. Archaeologists have tended to privilege the technological domain over social organization or religious and spiritual life, giving a primacy to stone tools. As one lithologist put it, "tools provide a thermometer for measuring intellectual heat." The tight fit between the long-dominant "man the hunter" hypothesis and the "man the toolmaker" paradigm has elevated technology to the element defining the prehistoric "Ages of Man": the Paleolithic and Neolithic Ages, the Bronze Age, the Iron Age.[29]

The privilege given tools (narrowly defined as finely articulated arrowheads, spears, axes, and the like) and the bones of the great beasts felled by them has been buttressed by disciplinary hierarchies that largely exclude women from fieldwork and cluster them in less prestigious fields, such as paleoethnobotany, zooarchaeology, museum work, laboratory analysis, and macro- and micro-wear lithic analysis. As in other big science, the archaeologist Joan Gero notes, women do not often direct "big digs," which involve procuring grants and permits, supervising large crews, acquiring and maintaining equipment, securing housing and subsistence for participants, and administering payrolls. Field-project directors may also help secure deep pit walls, stabilize grid systems, fashion ad hoc buildings for storage and lab space, improvise and maintain appropriate machinery, clear ground cover, move heavy stone, and backfill large pits. It is not that women are not capable, but that they are less likely to be chosen for jobs which require active, exploratory, out-of-doors, dominant, managerial, and risk-taking work.[30]

Sexual divisions in academic labor are especially evident in the high-status area of North American Paleo-Indian research, still widely known as "early man" studies, in which in the late 1980s men still carried out 90 percent of the fieldwork. Men in the field have defined as interesting only a small range of tools, namely elaborately produced fluted points (arrowheads, spearpoints, axes, and adzes) that are celebrated as exemplifying Paleolithic life and are typically interpreted as male innovations. The social value ascribed to projectile points in Paleo-Indian fieldwork is reinforced by gender segregation in lithic studies, in which men tend to dominate flint knapping. Knappers re-create ancient tools and actually use them to reenact ancient man's supposed activities—hunting, spear throwing, butchering—thus keeping attention tightly focused on hunting tools.

Gero challenges two aspects of this traditional "man the toolmaker" story. First, she points out that there is no evidence that women did not make these highly valued stone tools. Among some peoples, such as those who occupied the site Huaricoto in highland Peru between 200 and 600 A.D., there is evidence that they do. The association of men and tool production, Gero argues, rests on assumptions about divisions of labor between the sexes that are recent and particular to European and American culture. Second, she notes that there is good evidence that such a narrow definition of stone tools overlooks as much as 90 percent of prehistoric tool production. The emphasis on big-game hunting, which in fact probably provided only a small part of early humans' diets, has been considered one of the "great events" of prehistory, but overlooks the importance of flake stone tools (which may or may not have been used more commonly by women) and the work associated with them. Women archaeologists, who are well represented in lithic studies, typically study these flake stone tools and other informal instruments found on house floors, at base camps, and in village sites. These micro- and macro-wear studies, as they are called, focus not on re-creating past technologies but on determining how the stones were used in a wide range of activities, including nutting, leatherworking, grain harvesting, and woodworking. An expanded definition of tools (as Slocum argued some twenty years ago) opens up new questions about how the hunters' meat was prepared, what early people usually ate, and the economic and cultural goals of tool-making societies.[31]

Gender studies in archaeology are expansive. Feminist archaeologists commonly "find" and highlight the contributions of lost women in prehistory, as for example the innovations surrounding women potters. In the past archaeologists typically were interested in pottery only after mechanization (the development of the potter's wheel) and in its association with the development of commerce, the domain of men. Rita Wright asserts that the development of pottery (7000 B.C.) was an invention of major historic significance fostered by women.[32]

Feminist archaeologists also uncover gendered assumptions. For example, Conkey has noted that objects found in a grave are often assigned very different meanings depending on whether the grave is that of a woman or of a man. Pestles, for example, when buried with women are interpreted as mementos of women's grain-grinding activities, but when buried with men are thought to indicate that men manufactured them. The same is true of trade goods: when they are found interred with women, it is assumed that they were part of women's households; when they are found with men, it is assumed that men controlled trade. Similarly, atlatls (spear throwers) found with men are taken to symbolize hunting activities; when found with women they are thought to be purely ceremonial or related to the transfer of property.[33]

Finally, the archaeologists Patty Jo Watson and Mary Kennedy have revealed the power of the traditional paradigm, in which men are seen as the creators of culture, to obscure women's contributions to the invention of agriculture. Women everywhere, except in the contemporary industrial world, have been the primary cultivators of foodstuffs. Widespread consensus identifies women as gatherers and harvesters in prehistoric eastern woodlands, where agriculture in North America arose. The association between women and plants is firmly established in archaeology, anthropology, and evolutionary studies. Watson and Kennedy show, however, that when it comes to discussing the invention of agriculture, women disappear from the picture. Suddenly the coevolution of peoples and plants is seen as so "natural" (unintentional and automatic) that the plants seem "virtually to domesticate themselves."[34] Watson and Kennedy restore to women the role of active domesticators, whose agricultural feats provided staples—including sumpweeds, sunflowers, and chenopods (related to spinach and beets)—to the Late Archaic diet.

The philosopher of science Alison Wylie asks why, given its lateness, the feminist critique in archaeology emerged when it did in the late 1980s. Several "catalysts" set things in motion. The monolithic positivism in the field that Conkey referred to was shattered by a new attention to how knowledge "bears the marks of its makers." This opening allowed concerns about gender to be voiced. But more than theoretical or methodological shifts in the discipline, Wylie believes, sociopolitical events were responsible for the interest in women and gender as research subjects in the late 1980s. The second wave of the women's movement helped dramatically increase the numbers of women receiving degrees in archaeology; it also prompted people who may not have called themselves feminists to look more critically at equity issues and at how gender structured archaeological knowledge. Further, archaeologists were influenced by feminists in allied fields (in this case, cultural anthropology and history) and their successes in posing dazzlingly new research questions.[35]

Wylie also mentions the importance of people who organized influential conferences, such as the 1988 conference on Women and Production in Prehistory, that challenged scholars who had not considered applying feminist insights to their work to implement gender as a category of analysis. She mentions the Herculean labors of Gero and Conkey in this respect, but she might also have mentioned her own role as a leavening agent: her strong analytical reviews of the engendering of archaeology have served to consolidate feminist innovations in the field and to reveal the kind of intellectual ferment that can emerge from active collaboration between scientists and humanists. In the context of the current "science wars," it is important to point to productive working relations between philosophers, historians, and scientists where they do exist. Such collaborations in archaeology have sparked its practitioners, as Wylie reports, "to think differently about their discipline and their subject matter, to identify gaps in analysis, to question taken-for-granted assumptions about women and gender, and to envision a range of alternatives for inquiry and interpretation"—surely the very stuff of scientific creativity.[36]

8

Biology

GENDER analysis has made grand inroads into many areas of biology. While scholars have yet to study why the feminist critique has been so successful, the Boston University biologist Marian Lowe notes that in biology sex and gender were already important areas of study, that, as in medicine, many research areas have direct effects on women's lives, and that the relatively large number of women scientists has made possible a stronger feminist voice.[1]

Linguistic Decodings

A simple example of how gender has molded aspects of cell biology can be found in textbook accounts of conception, where the active sperm and the passive egg remained stock characters well into the 1970s. As the Swarthmore Biology and Gender Study Group and, more recently, the anthropologist Emily Martin have documented, in these sagas of conception the spermatic hero actively pursues the egg, surviving the hostile environment of the vagina and defeating his many rivals. The large and placid egg, like Sleeping Beauty, drifts unconsciously along the fallopian tube until awakened by a valiant sperm. The sperm penetrates the egg and conception is achieved.[2]

In 1983 Gerald and Heide Schatten highlighted efforts to revise fundamental notions of fertilization in an article appropriately entitled "The Energetic Egg." They portrayed the egg, like the sperm, as an active agent, directing the growth of microvilli (small finger-like projections on its surface) to capture and tether the sperm. Once the sperm is oriented in the right direction by the egg, its tail and digestive enzymes (some of which are activated by contact with the egg) allow it to enter the egg. The egg and sperm are portrayed as "partners"—perhaps a dual-career couple—working together toward successful fertilization. It is worth noting that the egg's cone of microvilli was documented as early as 1895 but was not considered worthy of research until some eighty years later.[3]

The account of the energetic egg has been hailed as an example of prejudice vanquished. Feminist critique is one among many ways of un-covering bias—as an additional experimental control to help scientists avoid error.[4] Moreover, thinking of the egg as an active partner has led researchers to discover previously unknown aspects of the egg's contri-butions to fertilization.

There is, however, another way to understand this new version of the story. We might also see it as a narrative of masculinization. Not only is the egg energized; it is masculinized, that is, ascribed the valued "active" characteristics of the sperm. Equality—this time for the egg—depends once again on reaffirming masculine values. Like women themselves, fe-male biology is here expected to assimilate the values of the dominant culture. Martin warns that as the egg becomes active or masculinized it is also seen as aggressive—a *femme fatale,* threatening to capture and victimize sperm: "New data did not lead scientists to eliminate gender stereotypes . . . Instead, scientists simply began to describe egg and sperm in different, but no less damaging, terms." The molecular biologist and professor of women's studies Bonnie Spanier interjects that in this instance the notion of equality between the contributions of the egg and the sperm is misleading, hiding the fact that the egg contributes more to biological reproduction than the sperm. The emphasis on "hereditary equality," she argues, diminishes the actual role of the egg as the larger gamete that contributes nutrients, organelles such as mitochondria and ribosomes, the cell membrane, and proteins crucial to the development of the zygote.[5]

Surely, one could object, science that bolsters gender equality is worthy of praise. If the new story of conception is still problematic, what is the correct reading? Many assume that stripping away gender bias allows scientists to see more clearly what is "really" going on in nature—to get closer to the truth free of gender. In this example, however, it is important to recognize that egg and sperm cells are still gendered. This time the "male" and "female" parts are portrayed as interactive partners, a view more in tune with currently fashionable human gender relations. We can-not free ourselves of cultural influence; we cannot think or act outside a culture. Language shapes even as it articulates thought. Gendering the egg and sperm places them in preexisting and complex sets of cultural meanings. By becoming aware of how we use language—by "waking up" metaphors, as Martin puts it—we can critically judge the imagery structuring our understandings of nature. A critical awareness of how gender influences science allows us to craft science according to preferred, rather than default, values. An awareness of culture can thus become a vital part of the research design. This improves our ability to understand nature and enables us to create better science.[6]

Even so, some critics of feminism see this kind of analysis as nothing more than "metaphor mongering." They object that scientists use metaphors in textbooks and other general materials in order to communicate more effectively with the public. In the world of research, no such language is used. How metaphors function in research environments is a question for ethnologists. Metaphors are not innocent literary devices used to spice up texts. Analogies and metaphors, the literary critic Susan Squier asserts, function to construct as well as describe—they have both a hypothesis-creating and a proof-making function.[7]

The story of the gendering of cellular nucleus and cytoplasm is an analogue to the story of the sperm and the egg: again, gendering things female led to the neglect of certain areas of research. Bonnie Spanier and Scott Gilbert tell how the rise of genetics—with its single-minded attention to the gene as the primary agent of life and the ultimate object of biological investigation—brought about an eclipse of embryology and the study of the cytoplasm in the 1950s. In this era the cell nucleus, the carrier of DNA, was seen as coterminous with the sperm, a masculine element carrying genetic information (a distant relative of Aristotle's "form"). The cytoplasm of the fertilized egg, the material body of the cell, was seen as receiving directions for development from the nucleus. The gendering of cells accounts at least in part for historical discounting of mitochondrial DNA and maternal RNA. Spanier notes a resurgent interest in cytoplasmic inheritance (known in the 1950s and 1960s as "maternal inheritance") and the study of mitochondrial DNA—known as the "Other" Human Genome Project.[8]

Gender was one reason for American geneticists' emphasis on the active intelligence of the nucleus over the passive body of the cytoplasm, but, as is usually the case, it was not the only contributing element. Disciplinary rivalry in the 1930s between American geneticists (who considered genetics superior to embryology and accorded the nuclear genome authority over cellular functions) and German embryologists (who did not make sharp distinctions between genetics and embryology, and many of whom accorded the cytoplasm an equal role in directing development) was bolstered by the fortunes of war. The association of genetics with Americans and of the study of embryology (and cytoplasm) with Europeans, especially Germans, aided the phenomenal growth of genetics in the postwar period as the scepter of scientific ascendancy passed to the United States.[9]

Linguistic decoding has also uncovered the influence of narratives of marriage and courtship (rituals remade by feminists in the nineteenth century and again the 1960s) in biology. At least since the eighteenth century, biologists have used marriage as an important heuristic for understanding couplings and reproduction in plants and animals. The great Swedish naturalist Carl Linnaeus, the celebrated "father" of modern taxonomies and nomenclatures, made the "marriages of the plants" the key to his renowned system of botanical taxonomy, known as the "Sexual System." Linnaeus not only identified the male and female parts of plants, he transformed them into marriage partners, configuring stamens as "husbands" *(andria)* and pistils as "wives" *(gynia)*. Since the 1860s the egg and the sperm have been called *gametes,* a term derived from the Greek *gamein* (to marry) to refer to a germ cell capable of fusing with another cell to form a new individual. More recently, as DNA has come to hold the position of "master molecule," other genes coding for enzymes running a cell's metabolism have been disparaged as "housekeeping" genes—invoking all the dreary chores associated with that term. These genes share the lowly status of poorly compensated housekeepers (traditionally wives), whose productivity rarely figures in a nation's GNP.[10]

Subjects rendered female—whether plant or animal—are often expected to conform to the demands of high femininity. Modesty became in the seventeenth century, and in many instances still is, a hallmark of Western feminine deportment. The first editor of the *Encyclopædia Britannica,* the eighteenth-century naturalist William Smellie, found this "distinguishing and attractive characteristic of the female sex . . . even so low [on the great chain of being] as the insect tribes." One of his French contemporaries, the botanist René-Louis Desfontaines, found it among the female parts of plants. He reported that while the stamens (the male parts) had visible orgasms, the pistils (the female parts) experienced little sexual excitement, "as if the law requiring a certain modesty of females were common to all organisms."[11]

Conjugal narratives need not necessarily follow Western laws and customs. Take the widespread practice in ethology of calling herds—of horses, antelope, elephant seals, and so on—"harems." The assumption is that one mighty male, acting as sultan, safeguards his females, who like a sultan's women reserve their sexual services for him alone. It has long been thought that horses, for example, run in harems. Recent DNA studies of mustangs show, however, that a given stallion typically sires fewer than a third of the foals in a band. In this case, as in so many others, the focus on male activity limits researchers' ability to "see" what lies outside the logic of the metaphor. Researchers who questioned the notion of a "harem" found that female mustangs range from band to band, often mating with a stallion of their choice.[12]

The application of narratives of courtship and marriage to plants and animals has forged a commitment to (hetero)sexual differences, even where sexuality is ambiguous or nonexistent. Biologists have chosen, for example, to understand reproduction in bacteria, such as E. coli, through the lens of sex rather than other possible optics. Roberta Bivins points out that the consequences of this go beyond the realm of language: experimental designs built on a model of bacterial conjugation (sex) were themselves first shaped by assumptions about sex and gender, and subsequently (because of their design) produced results which validated those assumptions. Thus, she concludes, "a scientific understanding of E. coli was constructed, both physically and linguistically, to gendered and sexed specifications."[13]

Bacteria were believed to be strictly asexual until the 1940s, when their "sex life" was first described in strongly heterosexual terms. Bacteria possess neither eggs nor sperm cells; indeed, in Lynn Margulis's words, "the extreme promiscuity of gene transfer in bacteria renders the idea of fixed sexes meaningless." Bacteria have, nonetheless, been defined as male or female based on the presence or absence of a "fertility" or F-factor (males are $F+$; females are $F-$). To transfer genetic material, the "donor" or "male" extends its *sex pili* to the "recipient" or "female." Unlike the case in higher organisms, the chromosomal transfer is unidirectional from male to female and the *male,* not the female, produces offspring. Further, when an $F+$ cell transfers a copy of its F-factor to an $F-$ partner, the recipient becomes male or $F+$. Because the donor cell replicates its F-factor during conjugation, it too remains $F+$. Thus all cells in mixed cultures rapidly become male ($F+$) donor cells: the females change into

males, the males remain male, and everyone is happy. A recombinant F −
(female) cell results only from a "disrupted" or failed transfer of DNA
(what Aristotle would have called an error of nature).[14]

Spanier comments that styling these bacterial interactions on heterosexual unions reinforces traditional notions of sexuality and gender and
underplays the transsexual implications of these transfers, in which sexual
encounters produce changes in sex. By 1990 much of this had been "corrected" in major textbooks; bacteria had ceased to be labeled male and
female. Evelyn Fox Keller notes, however, that the ascent of the single-
celled bacterium E. coli as the preferred research organism in the 1970s
was an additional factor suppressing interest in eggs, cytoplasm, and embryology.[15]

Another example highlights the far-reaching consequences of devaluing
things female or identified as female. Theories of mammalian sex determination tend to see the female as something incomplete or not fully
developed. The Brown University biologist Anne Fausto-Sterling has discussed the astonishing announcement by the MIT biologist David Page
in 1987 that human sex—that is to say, whether an embryo becomes a
female or a male—is determined by a master gene on the Y chromosome.
A latter-day Aristotelian, Page sees the female as lacking something (in
this instance an essential piece of the Y chromosome): "The female
WHT1013 carries 99.8 percent of the Y chromosome, she lacks only the
160 kb that comprises intervals 1A2 and 1B."[16]

Page's views are consonant with theories of mammalian sex determination that, until the mid-1980s, generally assumed that the male condition is actively produced by a gene-directed event, while the female
develops passively from the lack of intervention. As the story is traditionally told, mammalian embryos start out in an "indifferent" stage; they are
sexually ambiguous or bipotential. In humans, for example, the clitoris
and penis, and the labia majora and scrotal sac, are identical in the early
embryo. The Y chromosome is identified as actively determining sex. In
the course of time, the Y chromosome directs gene action that transforms
part of the "indifferent" gonad into the testis (the rest of the fetal gonad
withers away). In the absence of testosterone, the "indifferent" gonad
becomes the ovary. In that era and sometimes even today, fetal androgens
are also said to "masculinize" certain parts of the brain.[17]

The language that defines female development as an "absence" and
"lack" of the stuff needed to make a male would be inconsequential except
for its historical context. For thousands of years females have been

thought to be lesser or incomplete males, lacking some vital element (for Aristotle, heat; for Darwin, struggle for existence) required to propel them into the superior male state. In the 1970s the authors of a textbook on sex determination concluded their discussion by stating: "In all the systems we have considered, maleness means mastery; the Y-chromosome over the X, the medulla over the cortex, androgen over oestrogen. So physiologically speaking, there is no justification for believing in the equality of the sexes: *vive la différence!*"[18] This bald statement was removed for the 1982 edition of the same work.

In 1986 Eva Eicher and Linda Washburn, working with mice, attempted to overturn the traditional view of females as arrested in a more primitive or unfinished state. "Some investigators," they noted, "have over-emphasized the hypothesis that the Y chromosome is involved in testis determination by presenting the induction of testicular tissue as an active (gene-directed, dominant) event while presenting the induction of ovarian tissue as a passive (automatic) event. Certainly the induction of ovarian tissue is as much an active, genetically directed developmental process as is the induction of testicular tissue . . . Almost nothing has been written about genes involved in the induction of ovarian tissue from the undifferentiated gonad." This perspective has filtered into the literature. In 1992 David Page, while still focusing on the SRY (sex-determining region of the Y) gene, made a point of emphasizing that "a female is not a 'default' pathway" and that "the ovary is not the absence of a testis." "There are two alternative paths," Page now averred, "and following either of them is a very active process." According to this new analysis, in order to produce a male the SRY gene stimulates the production of hormones, one of which stimulates male development while the other inhibits female development.[19]

Even in this new era of gender sensitivity, residues remain of the sense of female "lack" and the focus on males that continues to see the Y chromosome as a crucial factor for determining sex in mammals. One might imagine an alternate description of the same phenomena. In medicine (see Chapter 6), the male body long was taken as the norm for medical research and the female body as a special case or a deviation from that norm. One might interpret sex determination from another perspective: rather than seeing the female as lacking something, one might see the female as the ground plan of the human body—the most fundamental state from which the male is a deviation or a less stable, and certainly in humans less long-lived, special case. The point, of course, is not now to privilege females over males, but to open up new perspectives by challenging ingrained assumptions.[20]

The figurative language and thought structures scientists employ can affect the content of science. The hypothesis-creating power given to heterosexuality, for example, has caused certain types of homosexual mating to be overlooked. The deliberate search for homosexual unions has brought to light thirteen species of whiptail lizards of the American Southwest, composed entirely of females, that can reproduce. Though an isolated female is capable of reproducing on her own, these lizards produce more eggs more frequently when they live together in pairs.[21] Again, the purpose of gender analysis is not to shift science away from politically incorrect metaphors and analogies toward politically correct ones but rather to uncover how the totems and taboos built into language influence the questions scientists might ask and the results they might obtain.

Gender as a Structuring Principle

Gender in biology goes beyond the assigning of masculinity and femininity to unsuspecting plants and animals. It can also become encoded in practices, institutions, and the research priorities of science. It is often assumed that innovation in science rests on the discovery of some greater truth. More realistically, there are many forks in the road to knowledge. Some paths are determined by the availability of funding, some by national emergencies or priorities, some by curiosity, and some by what Helen Longino has called "background assumptions."[22] Background assumptions are the "taken-for-granteds" that seem so innocuous that they become invisible to a professional community. These assumptions secure basic working practices, including a certain degree of consensus about problem definitions, the acceptability of solutions, appropriate techniques and instrumentation, acceptable jargon, and areas of imprecision and ignorance. These are protected and perpetuated by formal or informal exclusions from that community. In the absence of dissenting views, social values and practices often structure research programs in unconscious and unintentional ways.

Historical examples reveal how gender can become a silent organizer of scientific theories and practices, setting priorities and determining outcomes. Take again Carl Linnaeus's botanical taxonomy, an important precursor to modern systems. Despite the number and variety of botanical systems developed in the eighteenth century, Linnaeus's sexual system was widely adopted after 1737. We have already seen that Linnaeus built his influential system on the marriages of plants. He also used sexual dimorphism to divide the vegetable world (as he called it) into *classes* based on the "male parts" or stamens of the plants and *orders* based on the "female parts" or pistils.

Linnaeus focused his categories on important reproductive organs; his system did not, however, capture fundamental sexual functions. Rather it highlighted purely morphological features (the number and mode of union)—exactly those characteristics of the male and female organs which are least important for reproduction. In view of this fact, it is striking that Linnaeus also devised his system in such a way that the number of a plant's stamens determined the class to which it was assigned, while the number of its pistils determined its order. (In his taxonomic tree, class stands above order.) In so doing, Linnaeus gave male parts priority in determining the status of the organism in the plant kingdom.

A particular social structure—the legal subordination of women to men—seemed so natural to Linnaeus that he inadvertently made it an organizing principle of his botanical taxonomy. Culture can come to structure scientific theories and practices, usually not, as Longino has noted, as a willful imposition against the evidence, but through the way questions are posed and data interpreted.[23] Linnaeus was also a social conservative who wanted his daughters to grow up to be hearty, strong housekeepers, not "fashionable dolls" or bluestockings (as intellectual women in his day were sometimes called). Thus the notion that females (whether humans or gendered parts of plants) are naturally subordinate conformed to his political outlook and personal beliefs. His assumptions were sufficiently widespread that this aspect of his taxonomy passed unquestioned by most of his colleagues. Unquestioned background assumptions can perpetuate a self-reinforcing gender system. Linnaeus's plant taxonomy, subordinating female parts to male parts, reinforced and naturalized conventional social practices. Conventions of gender unconsciously attributed to nature, in turn, buttressed the disenfranchisement of women from the public culture of both science and civil society as social theorists attempted to build a just society based on natural law.

A final example from the eighteenth century illustrates the way gender assumptions may silently persuade scientists to choose one path over other equally valid ones. In 1758 Linnaeus introduced the term *Mammalia* into zoological taxonomy—a step that would be hailed in the twentieth century as the starting point of modern zoological nomenclature.

Linnaeus devised this term, meaning literally "of the breast," to distinguish the class of animals comprising humans, apes, ungulates, sloths, sea cows, elephants, bats, and all other organisms with hair, three ear bones, and a four-chambered heart. In so doing he idolized the female *mammae* as the icon of that class. His seemingly innocent choice of labels for the class of animals that unites humans with other primates held implications for hotly debated questions of his day: women's role in the state, their rights as wives and mothers, their access to education and the professions, and the structure of women's health care.

The question of how to place humans in nature—the question of all questions for the eighteenth century—led Linnaeus to abandon the term that for over two thousand years had served to denote what we now call mammals (along with most reptiles and several amphibians): *Quadrupedia*. In coining his new term *Mammalia,* Linnaeus did not draw from tradition but devised a wholly new term. As I have argued elsewhere, he could have derived a term from any of a number of equally distinctive, perhaps more universal, and certainly more gender-neutral characteristics of the animals he designated mammals. He might have chosen, for example, the term *Pilosa* (the hairy ones), *Aurecaviga* (the hollow-eared ones), or *Lactentia* (the sucking ones).

If Linnaeus had other valid choices, why did he focus on the maternal breast? His attention to the fully developed female breast had as much to do with unique qualities of the mammal as with eighteenth-century politics of wetnursing, maternal breastfeeding, and the contested role of women in science and the broader culture. Linnaeus's choice of the term *Mammalia* presented a problematic outcome for women. By emphasizing how natural it was for females—both human and nonhuman—to suckle and rear their own children, Linnaeus's work helped legitimize the restructuring of European society that was then under way. Linnaeus joined Enlightenment debates over child care, campaigning vigorously to abolish the ancient custom of wetnursing, the practice whereby elite women sent their children to the countryside to be nursed for a fee by peasant women (an often fatal alternative to maternal breastfeeding before the advent of baby bottles).

Opponents of wetnursing encouraged aristocratic and middle-class women to keep their children at home and care for them themselves, thus promoting the modern stay-at-home mom as a natural and appropriate social system of child care.[24] But this arrangement was not the only possible alternative to wetnursing. As in the case of scientific nomenclature, other alternatives existed in possible forms of social organization. Within early modern Europe, economic production and social reproduction took place side by side in guild households. Western culture might have found solutions other than the stringent break between public and private spheres that came to characterize life in the nineteenth century. Universities, factories, government buildings, and places of public work, debate, and gatherings might have created what we now call on-site daycare centers and lactating rooms, so that the process of reproduction might have been joined with production at the very dawn of industrial society and modern democratic orders. This, however, did not occur. Some years later, as revolutionaries in France discussed whether civic rights should be extended to women, the maternal breast—now raised from a natural phenomenon to a scientific taxon—also became a political entity, figuring in legislative debates about whether women would become citizens of the state. In the French Revolution, "the breasted ones" were not to be given public rights but encouraged to take up their "natural" duties in their homes.

Using more recent examples, Evelyn Fox Keller has documented how the masculine-identified public sphere and the feminine-identified private sphere have structured thinking in two areas of evolutionary biology: population genetics and mathematical ecology. Her concern is to show how the selection process that occurs in the context of discovery limits what we come to know. Keller argues that the assumption that the atomistic individual is the fundamental unit in nature has led population geneticists to omit sexual reproduction from their models. Though the critique of misplaced individualism is nothing new (Karl Marx pointed out that evolutionary theory embodied bourgeois notions of individuals and competition), the gender dynamics Keller reveals are. According to Keller, geneticists treat reproduction as if individuals reproduce themselves, effectively bypassing the complexities of sexual difference, the contingencies of mating, and fertilization. She likens the biologists' atomistic individual to the heuristic individual portrayed by mainstream Western political and economic theorists—both are "simultaneously divested of sex and invested with the attributes of the 'universal man' (as if equality can prevail only in the absence of sexual differentiation)."[25] Keller argues further that biologists use values ascribed to the public sphere of Western culture to depict relations between individuals (while values generally attributed to the private sphere to describe relations are confined to the interior of an individual organism).

Disciplines

We have looked at several tools of analysis—linguistic decodings that reveal how viewing organisms as male or female can lead to devaluing organisms or parts of organisms and to neglecting important areas of research; we have also seen that gender can set scientific priorities, silently structuring theories and practices. What has not been treated sufficiently in feminist scholarship is how the historical division of disciplines has influenced our knowledge of the world. Disciplines draw artificial boundaries within scholarship; they divide the world in sometimes arbitrary ways. As Ellen Messer-Davidow, David Shumway, and David Sylvan put it:

> For only two centuries, knowledge has assumed a disciplinary form; for less than one, it has been produced in academic institutions by professionally trained knowers. Yet we have come to see these circumstances as so natural that we tend to forget their historical novelty and fail to imagine how else we might produce and organize knowledge. Our world now seems so naturally divided into, say, biology, physics, sociology, and history that when we try to imagine alternatives to these disciplines, we think merely of combining them: biochemistry, sociolinguistics, ethnomusicology.

Disciplines set limits to what can and cannot be asked and by whom. They specify the objects we can study (genes, deviant persons, classic texts) and the relations among them (mutation, criminality, canonicity). They provide criteria for knowledge (truth, significance, impact) and methods (quantification, interpretation, analysis) and regulate access to the professions.[26]

The study of how inquiry is disciplined by disciplines adds a new dimension to feminist criticisms of the reductionistic treatment of living organisms in molecular biology. Feminists have converged in their criticisms of the Human Genome Project on director James Watson's notion that understanding the gene—the most "golden of all molecules"—and its sequences is the ultimate goal of biology. Evelyn Fox Keller and the biologists Ruth Hubbard and Anne Fausto-Sterling have traced the prominence of molecular biology to a radical redefining of "life" brought about by the influx of physicists into biology after World War II and their transfer to molecular biology of many principles dear to physicists—an emphasis on simplicity, for example, and the goal of reducing things to smaller and smaller units. Physicists, fresh from the Manhattan Project, imported to biology the attitude that mysteries can be solved. By reconfiguring life as the mechanism of genetic replication, they concluded that life itself was not complex, but alluringly simple. The dramatic "successes" of molecular biology were rooted in a process of bracketing, of making the problem manageable by focusing on the causal relations between identifiable and controllable elements. Other, less controllable processes (embryonic regulation, cleavage, gastrulation morphogenesis, and so forth) were set aside. Fausto-Sterling and Keller argue that success was achieved by the redefinition of what constitutes legitimate questions and adequate answers.[27]

A central criticism of the Human Genome Project is that it channels crucial resources into genetic research and away from other demanding projects. Ruth Hubbard contends that the "geneticization" of North America actually threatens health by diverting attention and resources from the poverty and malnutrition afflicting much of the world's population. Molecular biologists suggest that defects in genes cause disease and that knowing the exact location of each gene on the chromosome and its molecular structure is the first step toward an efficient cure of the disease. Genetic defects, however, are responsible for only a small percentage of diseases; the major diseases in the world are not genetic, but infectious. More technical machismo is not required, Hubbard maintains, to reduce (for example) U.S. infant mortality, the highest in the industrialized world. We need instead social and political programs that provide jobs, a basic standard of living, sex education, vaccinations, education about preventive medicine and healthy lifestyles, and prenatal and well-baby care.

Hilary Rose adds that the genome project undermines public health efforts that look outward to the context of everyday life to explain why humans get sick or stay well. The new genetics, by contrast, looks inward to the determining code, and while promises of improving health (curing cancer, for example) attract funding, the new genetics ignores many important variables of life, such as healthy public policy in transportation, food, agriculture, energy, and economics.[28]

Gender structures science at different levels: sometimes at the level of theories, sometimes in nomenclatures or taxonomies, sometimes in research priorities, sometimes in the subjects chosen for study. Those who have reservations concerning the Human Genome Project focus on a fundamental level, asking about priorities and outcomes of particular research projects. It is not desirable to cut off any line of human inquiry. Given a world of limited resources, however, difficult decisions must be made about which projects are to be pursued and which are not. In this context one must ask: Science for whom? Who stands to benefit in terms of wealth and well-being from a particular project, and who does not?

9

Physics and Math

Even some of the severest critics of studies of science—the physicist Alan Sokal, for example—are willing to acknowledge that examples abound of how gender has molded particular aspects of the life sciences. Many critics of feminism continue to claim, however, a certain purity for math and physics. The challenge goes something like this: Is there a concrete example of gender in the substance of physics or math? Can you point to gender distortion in Newton's laws or Einstein's theory of relativity? If not, the feminist critique is insignificant.

Can we, in fact, identify gender in the physical sciences as we have done in the life sciences? Does the fact that electrons do not have gender in the same way as some of the objects of inquiry in the life and social sciences make physics immune to feminist analysis?

Is Physics Hard?

What is it about physics that so vehemently excludes women? It seems odd that in the biological sciences (where, as we have seen, multiple negative understandings of females as passive or substandard abound) 38 percent of the Ph.D.'s are now awarded to women, while in physics, where far fewer examples of overt gendering have been discovered, only about 13 percent of new Ph.D.'s are women. In 1996 women constituted 3 percent of full professors of physics, 10 percent of associate professors, and 17 percent of assistant professors in Ph.D.-granting departments. In 1994 a full 36 percent of Ph.D.-granting departments had no women faculty members; among departments granting only bachelor's degrees, three-quarters had no women faculty members.[1]

This modern pattern belies women's long participation in the field. Laura Bassi, a physicist at the University of Bologna, was one of two or three women who held positions as professors in the eighteenth century (see Chapter 1). The French physicist Emilie du Châtelet was perhaps the most celebrated woman scientist of the eighteenth century. Her translation of Newton's *Principia mathematica* with a commentary (published after her death in childbirth) remains today the standard French translation of that work.[2] In the twentieth century Marie Curie, Lise Meitner, and Maria Goeppert Mayer all made major contributions, sometimes without the benefit of regular academic positions or even proper laboratories.

The very scarcity of women in physics may be insulating the discipline from feminist critique. There have been very few studies of gender in physics: Evelyn Fox Keller and Helen Longino, who in 1996 published a collection of "classics" in gender and science, named the physical sciences as one of two major areas in need of further work (the other was non-Western sciences).

Scholarship to date on gender in physics has followed several lines of investigation. Sandra Harding has questioned the prestige physics enjoys as the model science. Sharon Traweek and a number of women physicists have emphasized the noisy arrogance of the culture that tends to silence women (see Chapter 4). The physicist Karen Barad has identified a pedagogical style in physics that teaches students to value fun and irresponsibility over meaning and understanding. Others have emphasized how physicists' military ties have held women at a distance. Still others have analyzed how the fortress mentality of "value neutrality" has insulated the physical sciences from gender critique.[3] As noted in the case of archaeology, feminism has made its greatest impact in fields least anchored in positivist epistemologies, fields that have strong traditions of interpretative understanding, including critical and self-reflective thinking.[4] It is worth noting that the proportion of women in particular disciplines follows a hierarchy of perceived prestige of the disciplines, at least in U.S. universities and research communities.

One common explanation for the low numbers of women in physics is that physics is "hard." We are told repeatedly that the physical sciences are hard and that the life sciences, like the humanities and social sciences, are soft. It is possible to distinguish three different meanings of the supposed hardness of physics. First and foremost, the physical sciences are held to be epistemologically hard. As disciplines, they are considered mathematical, yielding "hard and fast" (also known as "robust") results, and grounded in stringently reproducible (to the eighth digit) fact, while the soft sciences and the humanities are characterized as having considerable breadth, permeable boundaries, and open-ended epistemological structure. In their ethos and telos, the so-called hard sciences are said to be "dispassionate," distant, abstract, and quantitative, while the soft sciences are considered "compassionate" and qualitative, perhaps introspective, and closer to everyday concerns.[5] Physics and the physical sciences are also supposed to be ontologically hard. They study hard, inanimate things—matter in motion—while the life sciences and humanities study soft, animate organisms—plants, animals, humans, and their behaviors. Finally, physics, chemistry, and the other physical sciences are seen as didactically hard, that is, difficult, requiring a high degree of abstract thinking, strong analytical skills, arduous work, and long hours.

The notion that the physical sciences are hard (in all three senses) emerged from a stringent brand of positivism in the early part of the century that has roots going back to the rise of British empiricism in the seventeenth century. Bertrand Russell wrote in the 1920s: "I mean by 'hard' data those which resist the solvent influence of critical reflection, and by 'soft' data those which, under the operation of this process, become to our minds more or less doubtful. The hardest of hard data are of two sorts: the particular facts of sense, and the general truths of logic." Doubt about these data, Russell stated, "would be pathological." Among hard data Russell also included facts of introspection, spatial and temporal relations, and some facts of comparison such as the likeness or unlikeness of two shades of color. Soft data included common beliefs, such as the belief in other people's minds, beliefs that require inference. Following from Russell's definition, the physical sciences are hard because they study things (facts of sense existing separately from us) and employ mathematics. Thus hardness and softness follow a continuum from the study of the external world, where little human inference and emotion are employed, to the study of the human condition and its products. Russell referred to Descartes in this regard, but was also reformulating distinctions made by the early empiricists (David Hume, John Locke, Bishop Berkeley, among others) between primary and secondary qualities. Primary qualities (matter, shape, and motion) were conceived as external to us and thus more "real" than secondary qualities (color, taste, smell) or things known to us through an admixture of human intellect.[6]

"Hardness" is thought to define a hierarchy of the sciences. According to this paradigm, hardness is determined by the degree to which the science is thought to be built on fundamental laws that describe reality. Physics ranks first. According to the Harvard physicist Gerald Holton, theoretical physics is the quest for a "Holy Grail," which is nothing less than "the mastery of the whole world of experience, by subsuming it under one unified theoretical structure." The biologist Scott Gilbert has suggested that modern academic disciplines follow a "Great Chain of Being" with the universe replaced by the university: "*Biology* deals with dirty matter: frogs, snails, puppy dogs' tails, blood, sweat, tears. *Chemistry* deals with matter purified and quantified: 2M H_2SO_4, 4 mg/ml KNO_3. *Physics* deals with idealized matter (when it deals with matter at all): ideal gases, electron probability clouds, frictionless surfaces. (If physics deals too much with material, it falls down a branch of the Chain to become engineering.) Finally, *mathematics* claims to have escaped matter altogether." Many physicists would probably be the first to agree that this hierarchy of the sciences also follows a scale of intelligence: physics is tough, hard, and analytical, not for the faint at heart. Its analytical methods and presumed ability to reduce complex phenomena to simple principles have been taken as the model to which all other sciences should aspire. Even the humanities went through a period of intense scientism in the 1970s, in which the goal was to quantify human endeavor to the greatest extent possible in order to arrive at greater certainty and institutional respect.[7]

The hardness of the science—in what it studies, how it studies it, and the degree of difficulty attributed to it—correlates with prestige, with funding, and, negatively, with the number of women in the field. The National Research Council has found that the more math that is required for a particular job, the higher the pay and the lower the rate of female participation. Conversely, the "softer" the science, the higher the rate of female participation (see Chapter 1). The elaborate gendering of disciplines has led Robert Westman to suggest that the history of science is "androgynous," combining the "hardness" of science with the "softness" of history. The imputed "hardness" of physics may not, however, explain the low numbers of women in the field: the gendering of physics as "hard," "analytical," and so forth is to some extent circular. Which came first, the few women in physics or the notion that it is hard and not welcoming to women? That physics is more difficult than other fields of study is part of its cultural image.[8]

The epistemological hardness of physics may be illusory—the result of narrowing the boundaries of investigation. The cosmologist Martin Rees has suggested that the question of the origins of the universe is "a grand problem but perhaps a more straightforward problem . . . and far easier than anything in the biological world." So while evolutionary geneticists are prone to suffer from "physics envy," it may turn out that biology is ultimately "harder" in the sense that the problems it undertakes encompass complexity not amenable to reduction to a few simple laws.[9]

As the physicist Karen Barad has pointed out, while Newtonian physics might be considered "hard" in a strictly positivist sense, quantum physics seems no "harder" than history or literary criticism considering that the phenomena labeled "elementary particles" depend on extensive instrumental and theoretical interpretation. The notion that physics yields certainty developed from Newtonian classical realism and its vision of a real world existing apart from us and knowable through objective inquiry. This notion of "objectivity" rests on a classical notion that physical properties are observer-independent attributes of objects. In quantum physics, by contrast, what are identified as properties of physical objects (positions and velocities of especially subatomic particles) cannot be attributed to either the object or the measuring instrument alone. The descriptive concepts of physics characterize our interaction with the world; they are not attributes of objects.[10]

The hardness of the physical sciences has been secured by the Cartesian clear and distinct separation of the practice of science from the critical examination of science. Barad sees "getting the numbers out," as the defining feature of contemporary physics and a uniquely American style of physics. She traces the development of this style to the 1920s and early 1930s when theoretical physics gained professional status in the United States: "As the center of physics shifted westward across the Atlantic, the disciplinary boundaries shifted as well: meaning, interpretation, and critical reflection were banished from the domain of physics." In the aftermath of the U.S. victory in World War II, this approach to physics became hegemonic around the world.[11] Questions of meaning, consequences, or social responsibility are not considered to be part of physics proper but to belong to other realms, such as philosophy, ethics, or history.

This may help explain the curious state of modern physics, which at the highest theoretical end couples unreflective materialism to a high-flying metaphysics. There are physicists who regularly see "the face of God" (George Smoot), seek "the God particle" (Leon Lederman), and strive "to understand the mind of God" (Stephen Hawking), thus endowing their quest with religious verve. Robert Wilson has remarked that "both cathedrals and accelerators are build at great expense as a matter of faith."[12] Yet the physicists' god is stripped of ethics and politics. God is "value neutral" in the same way they imagine their science to be. Consequently physicists can ascribe a higher meaning to their quest while still ignoring the social realities of their undertaking.

The "hardness" of physics does not, I think, fully explain the low numbers of women in the field. Sharon Traweek has shown that even though Japanese physics is modeled on the cooperative model of the extended household, women fare little better there than in the rampantly competitive U.S. physics communities. Traweek argues that one model for Japanese physics is the *ie,* or household, where individuals work, not for personal gain, but to maintain the household and its resources in order to pass them intact to the next generation. Decisions in the *ie* are made by consensus, a process Traweek characterizes as more democratic than the one used in the United States. In Japan women are criticized as being too competitive and individualistic, unable to work cooperatively, and not sufficiently nurturing to the newer group members. Traweek makes the interesting point that even though Western categories of gender are reversed in Japan—men are seen as cooperative and nurturing and women as individualist and competitive—women are as excluded from physics there as elsewhere. "There is nothing," she writes, "consistent cross-culturally in the content of the virtues associated with success. We do see that the virtues of success, whatever their content, are associated with men."[13]

Physics and the Military

The prestige physics enjoys has much to do with its success in war. (This is a prestige that may be waning with the end of the Cold War, the end of government-funded big physics—in the decision not to fund the Superconducting Supercollider—and the advent of the government-funded Human Genome Project which is quickly crowning molecular biology the premier science.) World War I was the chemists' war; World War II was the physicists' war. The historian of science Peter Galison has argued that "after the development of radar and nuclear weapons in World War II, science occupied an unparalleled position of prestige and power."[14]

Wartime science spawned what historians call "big science": large-scale science with multidisciplinary teams engaged in "mission-oriented" research working with capital-intensive equipment. Ties between science and industry characteristic of big science had already begun in the 1920s as physicists and engineers joined efforts to provide hydroelectric power in California, for example. The Manhattan Project represented big science at its apogee: a cooperative, nationally coordinated, government-funded research project involving thousands of the best researchers and directed toward the creation of a single product—an atomic bomb. The physicist Jerrold Zacharias said of this period: "World War II was in many ways a watershed for American science and scientists. It changed the nature of what it means to do science and radically altered the relationship between science and government, the military . . . and industry."[15]

By the 1950s the rapid growth of research and development funded by the military (though pursued chiefly in industrial and university laboratories) was of crucial importance for all those who worked in physics in America. In this period military R&D made up about 90 percent of all federal R&D; in 1986 military R&D continued at about 70 percent of all federal R&D. The physicist Paul Forman estimates that, in the 1980s, 55 percent of all American physicists and astronomers engaged in research and development activities worked on projects of direct military value.[16] As late as 1989, 27 percent of job-seeking physics graduates found work in the military (25 percent took jobs in manufacturing and 24 percent in service industries). In 1995 American universities received $1.3 billion from the Pentagon. In 1998 the United States had not yet achieved its goal of striking a fifty-fifty balance between military and civilian R&D funding. The end of the Cold War hit physics (and mathematics) hard, leading new Ph.D.'s to seek employment in nontraditional fields, such as finance, business, or occasionally even secondary school teaching.[17]

In the postwar period funds for what is called "basic, pure, or fundamental" research increased hand-in-hand with funding for applied research. Though insisting that the value of this research was not tied to its utility, Washington was clear that national security and economic strength rested on superior science. Military funding has shaped science by stimulating the growth of specific fields to the detriment of others. Graduate students in all fields tend to go where the money and jobs are. The Department of Defense's enormous financial resources led to the growth of materials science, cryptology, quantum electronics, and solid state physics, and artificial intelligence and neural networks within computer science.[18]

Is there something about the connection between the military and physics that has discouraged women's participation in physics? Feminist scholars have addressed this question in various ways. One approach has exposed the imagery of male pregnancy and birth surrounding the production and testing of the atomic and hydrogen bombs: the A-bomb was "Oppenheimer's baby," the H-bomb "Teller's baby." Successful bombs were male: "Fat Man" and "Little Boy." Carol Cohn in particular has revealed a world of defense intellectuals in which life and death were permuted, in which bombs became babies and creative people fathered weapons of mass destruction.[19] Cohn notes many reasons for the use of this and other highly sexualized imagery by defense professionals in the 1980s. One is to minimize the seriousness of war and its consequences: bombs viewed as babies seem less threatening. Another is that these images "suggest men's desire to appropriate from women the power of giving life." This type of argument too quickly throws out the baby of science with the bathwater of its military uses: it assumes that women—of all races, times, and cultures—are naturally peace loving, a proposition that does not hold historically.

The connection between physics and the military forged in World War II does shed some light on women's absence from the physical sciences. Early in the century women were often considered too frail to bear "the mental stress of hard study." As such they would hardly have been considered prime candidates for weapons research. While governments sometimes encourage women to enter science, as during the post-Sputnik years and in the 1980s (when the National Science Foundation's miscalculations of a shortage of scientists led to aggressive recruitment of women), this has not been the norm. Before the 1970s women who earned Ph.D.'s in science rarely found jobs in industry or federal scientific agencies. They were confined primarily to women's colleges, where there was virtually no government-funded R&D. Take the example of MIT, a place not known for its friendliness toward women. MIT emerged from World War II with a faculty twice as large as before the war, an overall budget four times as large, and a research budget ten times as large—85 percent of which came from the Atomic Energy Commission. At the end of the war the president of MIT stated: "The value [of MIT] to our country . . . is parallel with that of a fleet or an army." With no women on the faculty as late as 1960, women were not part of that convoy.[20]

The cultural conventions—variations on the notion that a woman's place is in the home—that have long prohibited women from joining defense activities have not been entirely successful. Lise Meitner, along with Otto Hahn, we should recall, discovered nuclear fission. Later in her life Meitner refused an invitation to work on the atomic bomb at Los Alamos. Although miserable in Stockholm, where she had fled when the Nazis overran Berlin, she declared, "I will not work on your bomb." After the war, despite being called "mother of the bomb," she continued to distance herself from the Manhattan Project and emphasized her opposition to weapons development. In an interview with Eleanor Roosevelt she underscored her opposition to war, stating that "women have a great responsibility and they are obliged to try so far as they can to prevent another war. I hope that the atom bomb not only ended this horrible war—here and in Japan—but that we will use this tremendous energy source for peaceful purposes."[21]

Of course there were other women who worked at Los Alamos in the 1940s, primarily as wives of the men building the bomb. Many of them ran schools, coordinated social events, had babies, cooked, cleaned, and created a somewhat tolerable life in the makeshift desert town. Those women spoke proudly of those years and "affectionately" dedicated their book, *Standing By and Making Do,* to their "husbands and to all the men who made the atomic bomb a reality." Other women, some married to men on the bomb project and some not, served as "computers" (calculating, before their namesakes, solutions to differential and integro-differential equations) at Los Alamos. Still others were scientists in their own right who contributed to the military effort. Across the country, some eighty-five women helped design and construct the atomic bomb. Leona Woods (later Marshall), part of Enrico Fermi's group at the University of Chicago, helped construct detectors for monitoring neutrons from the atomic "pile"—which became the first nuclear reactor. At Columbia University, Maria Goeppert Mayer performed theoretical studies on the thermodynamic properties of uranium hexafluoride and eventually won the Nobel Prize for her nuclear shell model. At Los Alamos, Elizabeth Riddle Graves helped determine what kind of neutron reflector should surround the core of the bomb. Jane Hamilton Hall, who worked as a senior supervisor for nuclear reactors under construction at the Hanford Engineering Works in the state of Washington, eventually became an associate director of Los Alamos National Laboratory. After the war many of these women dropped out of technical jobs.[22]

Women who worked on the Manhattan Project had very different reactions to the destructive force of the bomb. Joan Hinton became so repulsed by the militarization of American physics that she immigrated to China, where in the 1990s she still designed dairy farms. Jean Wood Fuller, in contrast, became an enthusiastic "female guinea pig" for the 1955 atomic bomb test in the Nevada desert. Relishing the blast at 3,500 yards from ground zero, she exclaimed that "women can stand the shock and strain of an atomic explosion just as well as men." Throughout the 1950s she devoted her energies to helping women prepare their homes for a nuclear attack.[23]

Today women still design nuclear bombs. In the late 1980s there were three female bomb designers at Los Alamos. One of them described her work as "being a peeping Tom on Mother Nature" (identifying herself, interestingly, as male in this sexually charged remark). For her a bomb was mainly a design challenge. A somewhat more circumspect woman works among the designers of nuclear warheads at Lawrence Livermore Laboratory. This woman of Japanese descent, whose aunt suffered severe radiation sickness at Hiroshima, justifiably fears nuclear weapons and is disturbed by certain aspects of American nuclear policy, such as the 1950s testing on Pacific Islanders. She defends her work on the grounds of her belief that nuclear weapons will never be used. For her the greatest threat is a nuclear accident, which she hopes to help prevent by perfecting the weapons.[24]

The anthropologist Hugh Gusterson has emphasized that weapons makers run the ordinary gamut of political affiliation, ranging from conservative to liberal, Republican to Democrat. It would be unfair to label them in any particular way. Mostly, however, the people Gusterson studied at Lawrence Livermore Laboratory do not think about or discuss politics. The socialization of scientists into the laboratory is "a process whereby political questions [are] transformed into technocratic questions."[25]

We return to the question why women are so poorly represented in physics and other physical sciences. Apparently not because it is harder conceptually, but rather because of its image, culture, associations, and organization. Many areas of physics during and after the war years became "big science." Women tend not to be in charge of "big science" just as they tend not to be in charge of large organizations such as the armed forces (Sheila Widnall, former Secretary of the Air Force, and Sara Lister, Assistant Secretary of Army Manpower and Reserve Affairs, are among the few) or Fortune 500 companies. Some fields of physics, such as high energy physics in which large accelerators are used, employ up to 500 Ph.D.'s on a single experiment. Big physics projects require teamwork along with what Lew Kowarski of the European Center for Nuclear Research has characterized as military-like hierarchies, autocratic leaders, committees, big money, and the participation of respected and strong personalities.[26] Women have not yet been considered prime candidates to direct these or other big science projects, such as archaeological digs (Chapter 7).

In addition to the question of women's participation in defense-related sciences or in big science, there are other questions about physics that are subject to feminist analysis, such as women's poor representation in theoretical physics—even though it does not depend on access to large pieces of equipment and the kind of organization this equipment breeds. The astrophysicist Andrea Dupree says that it is not the mathematics or physics that keeps women out of cutting-edge conjectural theory but that "extra bit of chutzpah, or aggressiveness or assertiveness." "To be a conjectural theorist," she continues, "requires a certain sense of inner strength, a certain sense of ego and the ability to be verbal, to be articulate, and to be aggressive . . . Theorists love to rank all the other theorists in the world." Women tend to choose problems whose solutions can be demonstrated more directly, perhaps because women have lower status in intellectual communities and their results tend to come under sharper scrutiny. Women often work on small-scale problems, like the surface of the sun, while men choose large-scale problems, like the structure of the universe, not because of inherent gender differences but because men are more likely to have the security and financing needed for large-scale problems, which may require ten to fifteen years to get results.[27]

Feminists are also asking about the relegation of applied physics to second-class status within the hierarchy of subfields, the structure of the physics community, how research groups are organized, how students are educated, how resources are allocated, what questions are considered important, and what answers are accepted.[28] The answers to these questions have a bearing on the content and character of the physical sciences.

In 1996 the unemployment rate for women Ph.D. physicists remained twice that of their male peers (3.8 percent compared with 1.9 percent) after controlling for job experience. As the MIT physicist Vera Kistia-kowsky has remarked, "Why would a woman want to get a Ph.D. in physics when she knows she can't get an interesting job and the pay is lousy?" Even in a field as "female-friendly" as medicine, a woman at the top of the profession remarked: "I have to be twice as smart and work three times as hard to get three-fourths the pay and one-half the credit."[29]

Math and the Female Brain

Almost half of the math majors in the United States are women, but only a quarter of the math Ph.D.'s, less than 10 percent of tenured faculty, and 5 percent of tenured professors in Ph.D.-granting departments. More tell-ingly, in 1992 women held only 5 of 288 tenured positions in the ten most prestigious math departments. Despite near equality at the undergraduate level, potent myths surrounding mathematical genius work to exclude women at the professional level. The mathematician Claudia Henrion has highlighted several of these myths. First, math is a field inhabited by rug-ged individuals who, working alone, create great mathematics by the sheer strength of their imaginative genius. Second, being a mathematician and being a woman are incompatible: math with its emphasis on mind is not a profession for the females of the species with their incommodious bodies that sometimes become pregnant and give birth. Third, mathe-matics provides certain, eternal, and universal knowledge arrived at through deductive reasoning and formal proofs.[30]

Henrion's vivid portrayal of gender in the professional world of math goes a long way toward explaining the unease many women feel. Little work has been done, however, on analyzing the content of mathematics from the point of view of gender; my review of the literature yielded but one example. The mathematicians Kenneth Bogart and Peter Doyle have suggested that certain problems have not been solved (or not easily solved) because of sexist assumptions. They cite the "ménage problem," first posed in 1891, which asks for the number M_n of ways of seating "at a circular table n married couples, husbands and wives alternating, so that no husband is next to his own wife." Bogart and Doyle suggest that only the tradition of seating one of the pair first—usually the wife "for cour-tesy's sake"—made this problem seem difficult and speculate that had it not been for this tradition the problem would have been solved fifty years earlier. The easiest solution requires that both be seated at once. (Bogart and Doyle do not comment on the highly Victorian and rigidly bourgeois character of the problem itself.)[31]

Some feminist critiques of mathematics have emphasized its limitations as a tool. Evelyn Fox Keller, for example, has emphasized that the availability of certain techniques and tools, such as highly developed mathematics, has pushed biology in certain directions to the exclusion of others. The notion of a single central governor, where fundamental characteristics of life derive from a single molecule (Watson's "master molecule"), she argues, has benefited from the fact that these models are more easily manipulated mathematically than models emphasizing global and functional interrelationships.[32]

There is in these critiques of reductionism nothing peculiar to women or to gender. Attempts to connect them to women are situated in an indefensible brand of difference feminism, such as Luce Irigaray's notion that the historical lag in elaborating a theory of fluids (in hydraulics) had to do with an association of fluidity with femininity.[33]

Let me delve here into but one of the debates especially pertinent to the question of women's advancement in science: women's mathematical ability. Math, as we have seen, serves as the critical filter for science careers. The prestige of a science often depends on its degree of mathematization, and the more math required for a particular job, the higher the pay and lower the rate of women's participation. It is popularly believed that boys are good at math while girls are skilled verbally. It is also popularly believed that these skills reflect innate sexual differences—that the differences we see in boys' and girls', men's and women's mathematics ability are a function of sex-specific brain organization.[34]

To what extent do men exceed women in mathematical ability? The German neurologist P. J. Möbius painted a bleak picture in 1900, estimating that it took one million women to find one with mathematical talent. Most women, he claimed, detest mathematics. Möbius was fond of saying that mathematics, which expresses masculine exactitude and clarity, stands in natural opposition to both "womanliness" and love: "A mathematical woman is an unnatural being, she is in a certain sense a hermaphrodite [*Zwitter*]." The great Swedish dramatist August Strindberg, opposing the appointment of Sofia Kovalevskaia as a professor of mathematics at the University of Stockholm in 1889, wrote: "As decidedly as that two and two make four, what a monstrosity is a woman who is a professor of mathematics, and how unnecessary, injurious and out of place she is."[35]

Today the answer to the question "Are men better than women at mathematics?" differs according to which measure one chooses. Standardized tests such as the Scholastic Aptitude Test (SAT), which are seen as measuring raw mathematical ability, favor boys; class grades, often dismissed as measuring mathematical achievement or learned skills, favor girls. Current orthodoxy holds that young boys and girls display few gender differences in mathematics. Differences begin to appear at age thirteen and grow throughout the high school years, with the starkest distinctions in mathematical and spatial ability appearing among high achievers. Nearly all sex-related differences are found among those scoring in the top 10–20 percent of students tested. For example, 8 percent of boys but only 4.5 percent of girls scored at the highest math levels on the National Assessment of Educational Progress (NAEP) test.[36]

Math is one area where naturists and nurturists continue to lock horns. There are a number of unresolved issues: Do gender differences in verbal and math ability actually exist or are they artifacts of the way tests are constructed and administered? Do gender differences in skill result from hard-wired brain structure? Or do they result from social experience, such as parents' and teachers' encouragement, courses taken, gendered stereotypes and expectations, and so forth?

Naturists offer a variety of biological explanations for what they take to be confirmed gender differences. One is the theory of greater male variability. Mathematical ability is taken to be genetic, carried on the X chromosome. Because a male inherits only one X chromosome, male intelligence is said to be highly variable. Female intelligence is considered less variable because a female inherits two X chromosomes, and the intelligence quotient contributed by one X chromosome may cancel out the intelligence quotient contributed by the other. Thus female intelligence, produced by two inherited chromosomes, hovers in a middle range, while male intelligence, unmediated by a second X chromosome, may be high, medium, or low.[37] There are at one and the same time more male geniuses and more male idiots.

A second explanation for greater male achievement in math has to do with degrees of brain lateralization. Studies of brain lateralization suggest that women do poorly in math because their brains are not as highly specialized as men's. Lateralization—the increasing specialization of the two hemispheres of the brain—continues until a child passes through puberty. Boys mature approximately two years later than girls, and thus are likely to have more highly lateralized brains with spatial and verbal functions located in separate hemispheres. (For right-handers, the left side of the brain specializes in verbal skills while the right side specializes in spatial skills.) Bilateralization, or lesser division between the left and right brain, in girls and women creates competition within the hemispheres, thus reducing spatial and mathematical ability. The "cognition crowding" hypothesis suggests that because women's verbal abilities are represented in both hemispheres, verbal processes tend to impinge upon neural space in the right hemisphere that in men is devoted more exclusively to spatial reasoning. Women derive certain benefits from their presumed bilateralization, the greatest being that they have lower incidence of aphasia, or speech disorders, after damage to the left hemisphere.[38]

Brain research has emerged as a hot new field, pushed by new technologies such as functional magnetic resonance imaging and positron emission tomography (PET) that measure changes in cerebral blood flow, allowing researchers to identify more exactly the location of specific brain functions. The neurologist Richard Haier recently PET-scanned male and female students as they solved SAT math problems and found that they used their brains very differently in this regard. High-scoring men (SAT scores of 700 or above) used their temporal lobes intensively—more than either low-scoring men (scores of 540 or so) or high-scoring women. High-scoring women showed no difference in brain activity from low-scoring women, suggesting that the high-scoring men's achievement was associated with effort.[39] The high-scoring men and women performed equally well. Nonetheless, they seem to use their brains differently.

Nurturists offer markedly different explanations for boys' domination of the upper-level test scores. A common one is that a larger percentage of boys than girls take the highest-level math courses offered in high school. A more controversial explanation is that girls tend to employ conventional strategies in solving problems, things they learned in high school, while boys use unconventional strategies, making boys more independent and successful on current tests.[40] Girls' aversion to risk or unwillingness to engage in unconventional problem solving correlates with studies reporting lower self-confidence among young women. Naturists suggest that different approaches to problem solving between the sexes reflect brain organization. Because of girls' brain bilateralization, their strong verbal abilities may prompt them to use verbal cognitive style when solving spatial problems.

The most challenging explanation today is that mathematical aptitude tests are biased. Naturists tend to assume that the SAT is a neutral instrument. But do current tests measure native ability as advertised, or do they favor young men? Take the example of the SAT, prepared by the Educational Testing Service in Princeton, New Jersey, and taken by 1.5

million sixteen-to-eighteen-year-olds annually. The purpose of the test is to predict first-year college performance. As every hopeful high school student knows, the stakes are high. Top scores are required to enter the most prestigious colleges and universities and to receive the best scholarships.

The SAT has two parts: the verbal and the mathematical. Despite the fact that cognitive studies generally show that girls are more verbal than boys, significant gender differences do not show up on the verbal portion of the SAT. Currently boys outscore girls by about 10 points (considered statistically insignificant). This was not always so. Before 1972 girls outscored boys, and they still score higher than boys on the verbal sections of two other major surveys: the NAEP and the National Educational Longitudinal Survey. What happened with the SAT? It has been recognized since 1942 that "intellect can be defined and measured in such a manner as to make either sex appear superior" and that conflicting data regarding sex differences in mental ability "must be attributed to differences in tests." The original Binet test of 1903 showed girls to be more intelligent than boys according to its measures. Binet fiddled with the test until both sexes tested equally. As Phyllis Rosser of the Center for Women Policy Studies has documented, in the early 1970s the Educational Testing Service set out to make the SAT-Verbal more "sex-neutral." Its efforts resulted in a shift of 3–10 points from girls to boys—a result that ETS considered gender neutral, though one that in fact favored boys slightly.[41]

The male advantage was achieved by increases in science and sports content in the reading-comprehension passages. On the November 1987 test 66 percent more boys than girls answered the following question correctly:

Although the undefeated visitors _____ triumphed over their underdog opponents, the game was hardly the _____ sportswriters had predicted.
   A. fortunately . . . . . upset
   B. unexpectedly . . . . . classic
   C. finally . . . . . rout
   D. easily . . . . . stalemate
   E. utterly . . . . . mismatch.

Generally boys outperform girls on questions related to sports, science, or business, and on questions dealing with concrete information. Girls outperform boys on questions relating to aesthetics, philosophy, human relationships, and on questions using abstract concepts and ideas.[42]

ETS has made no comparable effort to balance the SAT-Math, on which boys outscore girls by between 41 and 52 points, or one-half of a standard deviation. The gender gap in math scores has persisted since 1967, when data on sex differences were first collected. Women's scores have not risen despite the increased number of math and science courses they now take.[43]

There is good evidence that the SAT-Math could be manipulated to decrease the current difference between boys' and girls' scores. The psychologists Elizabeth Fennema, Janet Hyde, and Susan Lamon argue that the math gap between males and females is narrowing, though this change is not reflected in SAT scores. As early as 1973 Thomas Donlon of ETS noted that the gender gap on the SAT-Math could be reduced by an increase in the number of algebra questions (on which women excel) and a decrease in the number of geometry questions (on which men score better). A study of the November 1987 SAT-Math confirmed this finding and suggested that the content of verbal problems can favor one sex over the other. Students tend to skip questions with unfamiliar content, and girls typically complete fewer problems than boys. On the 1987 test boys outscored girls by the widest margin on a question having to do with basketball team statistics. Finally, the current format of the test—timed and multiple-choice—can influence boys' and girls' performance. Girls tend to score higher on essay and open-ended questions; they also do well on contextual questions such as those asking about the amount and type of information needed to solve a problem. Girls tend to react badly to time pressure. As critics of the test have pointed out, it is not clear that emphasizing speed—requiring snap judgments rather than analysis and reflection—tests the most important aspects of intellect. Girls are also less likely than boys to risk guessing at the right answer. Girls' scores improved dramatically when testmakers removed the "I don't know" option from the NAEP, forcing girls to guess when they did not know an answer.[44]

Considering the gender bias built into it, how useful is the SAT? Its purpose is first and foremost to predict grades for the first year of college. As mentioned in Chapter 2, the SAT tends to underpredict women's grades and overpredict men's. A study of 4,000 Maryland high school students, for example, found that girls who earned higher grades than boys in pre-calculus and calculus classes scored significantly lower (37–47 points) than the boys on the SAT-Math. The ETS's own studies indicate that women do as well in college math courses as men with significantly higher scores on the math SAT. Hyde, Fennema, and Lamon also found that the SAT showed larger sex differences in math than any of the other college-admission tests. (On the NAEP, for example, in 1992 boys outperformed girls in math by only a small margin.) In light of these findings, Federal District Judge John M. Walker ruled in 1989 that the SAT discriminates against girls, and prohibited the New York State Department of Education from using SAT scores as the sole basis for awarding merit scholarships. In the late 1980s MIT also took steps to counterbalance the apparent bias in the SAT by admitting students, especially girls with good math preparation, who scored under 750 on the SAT-Math.[45]

In her study of the SAT Phyllis Rosser found that largest gender disparities between test scores and academic performance occurred among boys and girls with the highest grade point averages (A-plus to A). Girls receive 5 percent more A-pluses than boys in subjects relating to verbal skills and 10 percent more A-pluses than boys in math classes. Yet these girls score significantly lower on the SAT than boys with equivalent GPAs. This means that "the highest achieving girls are penalized the most by the SAT gender gap." These girls, who on the basis of their grades might have been accepted at prestigious colleges and won distinguished scholarships, are often disqualified by their test scores. Scholarships awarded using test scores alone are twice as likely to go to boys as to girls. Low scores on standardized tests can also exclude girls early on from academic enrichment programs and accelerated courses, including programs for the "gifted and talented." Lower test scores also tend to lower women's academic aspirations as well as their perceptions of their own abilities. Women often apply to less prestigious colleges than their grades would support.[46]

One might argue that grades and aptitude tests measure different skills. Grades may evaluate a variety of qualities—neatness, diligence, ability to complete work or follow directions, improvement over time—in addition to mastery of the material. Teachers may factor in social skills such as "good citizenship." Standardized tests, in contrast, evaluate a smaller range of skills, such as analytical reasoning and the ability to work under pressure. It is not clear, however, that the latter skills are the most important for long-term success or scientific creativity. Perhaps the most telling finding in this area is that ability as measured on standardized tests is not closely related to research performance in science.[47]

In mathematics as in many other fields, few efforts have been made to study gender differences in relation to other important variables, such as ethnicity, culture, or class. If for a moment we assume that standardized tests do accurately measure a difference in mathematical ability between boys and girls in the United States, is this difference consistent across cultures and across time? Naturists, such as Camilla Benbow and Julian Stanley, argue that it is. They see the superior male mathematical abilities—quantitative and spatial abilities as well as field articulation—as hard-wired in the male brain. In order to test the fixity of gender differences in mathematical ability, Benbow and Stanley had the U.S. SAT-Math translated into German and Mandarin Chinese and administered to students in Germany and China. Their results showed the same range of sex differences in these radically different cultures, leading them to conclude that indeed "sex differences may partly be biologically induced." As biological factors, they suggest greater brain lateralization and exposure to high levels of testosterone that slow the development of the left hemisphere and thereby enhance the development of the right hemisphere (where spatial abilities are located). Whatever male achievement may be in the United States, American students—neither boys nor girls—do not do well by world standards. In 1989 U.S. thirteen-year-olds placed ninth among twelve nations in science skills.[48]

Studies of boys and girls from different ethnic groups within the United States show some surprising results. Girls in Hawaiian public schools, for example, outperform boys both in the classroom and on standardized tests, especially among Filipino, Hawaiian, and Japanese populations. Differences are found as early as the fourth grade and increase as students mature. Other studies have suggested that African-American and Hispanic high school girls test higher than boys of those ethnicities in mathematical ability. It should also be pointed out that Asian-American boys outscore European-American boys by 26 points on the SAT-Math, and that European-American boys average only 14 points higher than Asian-American girls (not considered statistically significant). The few comparative studies of mathematical ability that have been done suggest that sex differences in mathematical achievement vary by ethnicity along a continuum ranging from moderate differences favoring girls to large differences favoring boys.[49]

Class can also affect gender differences in scores on the SAT-Math. It has long been known that the SAT test scores correlate highly with family income and tend to reflect class and educational advantages. But the correlation between class standing and test scores is highest for boys. Girls at every income level score lower than boys with comparable family incomes.[50]

It is generally assumed that high mathematical ability is crucial for success or even interest in science. Indeed, as the math content of a science increases, the number of women in that science decreases. Although facility in mathematics is undoubtedly necessary for most scientific fields, the direct relationship between mathematical ability and success in science has yet to be explored.[51] A U.S. Department of Education study showed that when math scores were the same, nearly twice as many men as women pursued physics. It is not, then, only a lack of ability that is keeping women out of science; something else is producing the disparities in men's and women's participation in academic mathematics.

The question of gender in the content of physics and math is complicated and requires further investigation. This is a task for the best physicists, philosophers, and historians of science with rigorous training in gender studies of science. Physics has been insulated from gender critiques partly because so few people are trained to undertake them. Members of a new generation of physicists, however, either have training in gender studies or are actively seeking to collaborate with those who do.

Empirical study may reveal that gender does not permeate the most abstract level of human endeavor. It does not necessarily follow, however (as some would have it), that the feminist enterprise stands or falls on finding such examples. What has been demonstrated is that gender abounds in the cultures of math and physics, determining to a certain extent who gets educated, gets funded, enjoys prestige, and can build upon opportunities. The content of physics is not distinct from its cultures; cultures—shared beliefs, expectations, "taken-for-granteds," and material well-being—mold many aspects of the various sciences. The greatest physicists have been those who have asked the right questions. Newton asked why the moon fell (when everyone else assumed it did not); Einstein asked what the world would look like if you rode along with a beam of light.[52] Ultimately, the culture of physics sets conditions for who has the training and the opportunity to ask questions. Feminism has made significant contributions by asking new questions, questions that often stand at odds with the foundational assumptions in a discipline. It remains to be seen what these questions may be in the fields of physics and math. Getting the right answers—turning the crank— may be gender free. But it is often in setting priorities about what will and what will not be known that gender has an impact on science. It is also perhaps here that the greatest feminist contributions will be made.

Conclusion

HAS feminism changed science? Since the 1950s, when the scientist was popularly conceived as a lone male genius peering into a test tube, and the 1960s, when the "future elite of science" was conceived as all boys (with fiery red hair), expectations about who will become scientists have undergone a sea change (Figures 8 and 9). More women now run government agencies, preside over university departments, and hold prestigious academic chairs. Interest in monitoring the situation has prompted the U.S. government, since 1982, to publish a biannual report on women's standing in science. Seen from a historical perspective, women's ascent has been remarkable. Progress, of course, is never inevitable, never guaranteed. In physics women's numbers have not budged significantly in a decade, and their status may even have declined from the late eighteenth century when Laura Bassi delivered her lectures at the University of Bologna.

More important, feminism has in many instances changed the content of human knowledge. Primatologists no longer see nonhuman primate society exclusively in terms of aggressive and territorial males. Archaeologists now conceive of "first tools" in terms of digging sticks, baskets (used for gathering), and slings (for carrying babies) as well as the traditional hunting instruments—elaborately produced arrowheads, spearpoints, axes, and adzes. Biologists no longer talk about fetal androgens as "masculinizing" certain parts of the brain. Federal law requires medical researchers to test procedures or drugs on a proper mix of women and men. Feminist influence has not been felt uniformly across the sciences. In physics and math, we wait for the people with the proper training and

Profile of
a New Elite

Figure 8. Profile of a new elite: 1964. Source: Margenau et al., eds., *The Scientist.* © Myron Davis. Reprinted with permission.

opportunity to explore the impact of gender on those bodies of knowledge.[1]

How do we proceed from here? How do we continue to turn a critical understanding of women's historical relationship to science into productive cultural change?

Figure 9. Profile of "women" (here European-American and Asian-American girls) in science, 1993. Source: *Science* 260 (16 April 1993). © Sam Ogden. Reprinted with permission.

The Academy

Feminists often set particular goals for science and define this as "feminist science." There is no shortage of speculation concerning ideal stances to take. In 1983 Peggy McIntosh identified rites of passage through distinct levels of understanding from a primitive "womanless science," to the liberal approach of adding women to "science as usual," to a difference-feminist approach of looking at things from the "female point of view." (The final stage called for reconstructing research and the curriculum in a rather vague way "to include us all.") Carolyn Merchant calls for a "partnership ethic," and I myself for a "sustainable science." Hilary Rose urges science practitioners to engage in equal measure "hand, brain, and heart."[2] Linda Fedigan has urged that primatology—with its distinguishing characteristics of humanitarian, ecological responsibility, reflexivity, and gender equality—is a feminist science (see Chapter 7). Merchant, Fedigan, and I (in my notion of sustainable science) focus on the values driving scientific research. The problem with defining a "feminist science" according to a set of values is that terms such as cooperative, interactionist, holistic can mean different things to different people and in different historical contexts. Even should some community of feminists reach an international accord on "feminist science," it would be difficult to implement a particular set of ideals within current science departments and funding agencies.

Donna Haraway and Sandra Harding have taken a slightly different tack, calling for adding an understanding of social context to scientific research (Haraway's "situated knowledge" and Harding's "strong objectivity"). While they advocate analysis rather than goal-specific values, it is no easier to integrate situated feminist knowledge or strong objectivity into science than to integrate any more specific feminist value, such as cooperation. As Robert Proctor has noted, military research—on the atomic bomb and artificial intelligence, for example—is highly situated, self-conscious knowledge, but hardly feminist.[3]

The desire to create a "feminist measuring stick" that will tell us when a science is feminist does not sufficiently allow for shifts in feminist theory and practice. The goal is not to create a feminist science, if that means (as it does for many critics) a special or separate science for women or feminists. Science is a human endeavor; it must serve us all, including women and feminists.

What is needed at this time is history, philosophy, and theory of science that analyze specific examples of gender in science—of the sort I highlighted in earlier chapters. What we need is a healthy working relationship between scholars involved in developing gender critiques of science and those doing science. In the fields where gender analysis has been most influential—medicine, primatology, biology, and now archaeology—it has been a highly collaborative effort. As we have seen, reforms in biomedical research at the National Institutes of Health (NIH) required the joint efforts of academic feminists, congressional leaders, medical doctors at NIH, and a healthy women's movement. In some areas, such as archaeology and biology, the collaboration has taken place within academia, where humanists and scientists have worked productively across C. P. Snow's two cultures. In other instances the feminist and the scientist have been one and the same person: certain anthropologists and primatologists (Marilyn Strathern, Sherry Ortner, Linda Fedigan, Adrienne Zihlman) number among major feminist theorists. It would be a misconception to think that feminism is something imposed upon science from the outside.

How can gender analysis be activated in other sciences, especially physics, chemistry, mathematics, and computer science? In physics the National Science Foundation has, since 1990, sponsored "site visits" to improve the climate for women. Initiated by prominent physicists like Mildred Dresselhaus and Bunny Clark, these visits aim at increasing the number of women in physics by actively recruiting women students and faculty, inviting women to speak at colloquia, and so forth.[4] One could imagine adding to these site visits (now devoted to women's career development and to making physics departments more friendly toward women) a robust analysis of gender dynamics in the content of the science, its research priorities and directions.

Another way to integrate a critical understanding of gender into science would be to have science students take courses on the history of gender in science.[5] Only in the past twenty years have such courses been available. Science students, however, may be told they do not have time to take these courses. To ameliorate this situation, a number of universities have history of science courses structured into the science curriculum. Stanford University's Values, Technology, Science, and Society program until recently was responsible for a special section of the university's world civilization course focused on science and technology. The University of Minnesota is exemplary in hiring and tenuring history of science faculty members within science departments, not in a special division of history where they would have little day-to-day contact with science colleagues. The Minnesota faculty offers to science students a variety of courses, ranging from the history of ancient science to the history of computing and engineering ethics. Schools of medicine have long housed their anthropologists, ethicists, and historians internally. Courses on gender and science offered in any of these contexts have the potential to provide students with both a historical understanding of women in science and the tools of gender analysis that can open up new vistas for future research.

Gender analysis can also become part of standard science courses. Curricular reform in science has been popular in the last few years and has yielded new approaches, such as chemistry in context, guided design calculus, hands-on laboratory experiences, exercises in collaborative learning, and emphasis on practical applications. Some of these classes also incorporate materials on gender: Scott Gilbert's Biology and Gender Study Group (a group of his students at Swarthmore College) wrote "The Importance of Feminist Critique for Contemporary Cell Biology"; Gilbert reports discussing and debating feminist materials as a regular part of his lab classes. Gilbert has also written an influential textbook, *Developmental Biology*, that integrates the new findings on gender into mainstream science. This provides an immediate and powerful corrective. The host of students who learn college biology from this text—future scientists and doctors, or future humanists—are also given critical tools for recognizing gender bias in biology.[6]

Tools of Gender Analysis

Furthering feminist research in science requires smart analytics. Gender analysis should act as does any other experimental control to provide critical rigor; to ignore it is to ignore a possible source of error in past and also future science.[7] Tools for gender analysis are as diverse as the variants of feminism and of science. As with any set of tools, new ones can be fashioned and others discarded as circumstances change. Not all analytical devices are peculiar to feminist studies—some are simply good history, sharp critical thinking, good biology, precise use of language. Some transfer easily from science to science, others do not. A number of analytics for designing woman-friendly research emerge from the examples of gendered science we looked at in preceding chapters.

Analyze Priorities and Outcomes

Feminism has made its greatest contributions by asking new questions, questions often at odds with fundamental assumptions in a discipline. One of the most important gender analytics looks at scientific priorities. How are choices made about what we want to know (and what we choose not to know) in the context of limited resources? And about who benefits in terms of wealth and well-being and who does not from a particular research project? Political interests and funding decisions spotlight certain portions of nature that become known while others are neglected. A good example for both its strengths and weaknesses, as we have seen, is research on women's health. Improving health care for women has not required new technical breakthroughs: it has required new judgments about women's social worth and a new willingness to invest in women's health and well-being.

Analyze Subjects Chosen for Study

Tools of gender analysis often have the virtue of addressing questions concerning both *women* in science (their position in the scientific community) and *gender* in science (how gender influences content). Analysis of the sexual composition of groups, for instance, can apply to constituting a search committee, achieving sexual equality at a conference, as well as structuring a representative sampling of animals or humans for a particular experiment or set of observations. The choice of study subjects can also have implications beyond those relating directly to females and males. Linda Fedigan has discussed the "baboonization" of primatology in the 1950s, when savannah baboons, one of the most aggressive and male-dominated of all primates, became the preferred model for ancestral human populations.[8] In this instance the choice of model subject introduced a potent antifeminist element.

Analyze Institutional Arrangements

Much of the gender analysis of science looks directly at the content of the sciences. It is equally important to scrutinize how institutional arrangements—whether these be informal "invisible colleges," rigorously formalized universities, scientific societies, or modern laboratories—structure the knowledge that issues from them. Gender becomes an important element where there is a strong relationship between the prestige of scientific institutions and women's standing within those institutions. We saw in Chapter 1 that women's participation declined as informal arrangements gave way to the professionalization of science in the latter part of the eighteenth century. We also saw that women's fortunes within modern universities waxed and waned according to the fortunes of war (their numbers increased during World War II but fell when men returned after the war) and of national legislation (women made tremendous gains when sexual discrimination was made illegal). Gender hierarchies also order women within the disciplines (as discussed in Chapter 9). It should give us pause, as we consider the relative value attached to various disciplines, that many modern disciplines had their origins in the German university system, from which women and their concerns were stringently excluded.[9]

Analyze the Cultures of Science and Domesticity

Tools of analysis have also brought to light gender dynamics in the cultures of the sciences. The distinguishing marks of a successful professor of English are not the same as those of a professor of physics, nor are the distinguishing marks of a successful female professor of physics necessarily the same as those of a successful male. The culture keeps members in line, quietly governing their dress, speech, and general deportment. Beyond regulating the behaviors of their practitioners, cultures foster intellectual styles that guide research programs. Helen Longino has discussed the way communities of researchers form "background assumptions"— the givens that serve as the bases for mutual understanding and effective research.[10]

Similarly, attention should be given to the relationships between science and domestic arrangements and the extent to which the former relies in unstated and often invisible ways on the latter. As we saw in Chapter 5, domestic arrangements are part of the culture of science.

Decode Language and Iconographic Representation

Language builds coherence within scientific cultures, and much gender analysis has focused on the rhetoric of scientific texts and images. Gender stereotypes are not innocent literary devices used to abbreviate thought. Analogies and metaphors construct as well as describe—they have both a hypothesis-creating and a proof-making function in science. They can determine the direction of scientific practice, the questions asked, the results obtained, and the interpretations deduced. Fundamental concepts in any field should not be taken for granted but should be set within historical frameworks of meaning. As Evelyn Fox Keller has emphasized, "sharing a language means sharing a conceptual universe" within which assumptions, judgments, and interpretations of data can be said to "make sense." Gendering the egg as passive and sperm as active, for example, places them within a deep matrix of cultural and historical meanings.[11]

Refurbish Theoretical Frames

There has been controversy about how deep gender analysis goes and whether feminists have contributed to the remaking of basic theoretical understandings of their disciplines. Feminists working in evolutionary theory, for example, have been criticized for merely adding females to standard theoretical frameworks. At its best, gender analysis interrogates what needs explanation and what counts as evidence. The archaeologists Margaret Conkey and Joan Gero have noted that male-identified stone tools figure among highly prized data tracing the "progress of humankind" (see Chapter 7). These potent symbols of "early man" tend to obscure other aspects of prehistoric life, such as nutting, leatherworking, grain harvesting, and woodworking—all of which were done using non-standardized stone tools.[12]

The philosopher Elisabeth Lloyd has revealed a different example from evolutionary theory. Lloyd, following Richard Lewontin, Stephen Jay Gould, and others, has questioned the primacy given adaptation in animal evolution. The propensity to link sexual activity narrowly to reproduction, she argues, has resulted in misleading accounts of autonomous female sexuality, especially female orgasm. According to Lloyd, women are (mistakenly) presumed regularly to experience orgasm with intercourse, as men do; Lloyd suggests that as many as a third of orgasmic women never have orgasms during intercourse. Women are further presumed to return to the resting state following orgasm, as men do. Female orgasm, Lloyd argues, is not tied to reproduction but, like male nipples, results from homologous embryological structures in males and females. Because orgasm is strongly selected for in males, females are born with the potential for orgasms. Lloyd concludes that extrapolating male models to females has led to misunderstanding of females and their role in evolution.[13]

Martha McCaughey further questions heterosexualist background assumptions that are often present in evolutionary theory. Again focusing on the asymmetries between male and female orgasm, she suggests that the disjunction between female orgasm and reproduction may have adaptive purposes. Female bisexuality, she proposes, may result from evolutionary advantages deriving from spaced pregnancies.[14]

Reconsider Definitions of Science

Finally, gender analysis has challenged what counts as science. Voltaire's proclamation of 1764 that "all the arts have been invented by man, not by woman" was echoed in 1991 when Stephen Cole and Robert Fiorentine asserted: "Women have achieved less than men in science. This statement is true no matter how we choose to measure achievement." Exploration of what is considered science—using ethnographic tools—can also influence the evaluation of women's contributions. Ellen Messer-Davidow, David Shumway, and David Sylvan have shown that what counts as science results in part from disciplines producing their own "economies of value," manufacturing their own discourse, regulating jobs, allocating funding, conferring and guarding prestige. Much that has not been counted as science has come from or treated the private side of life and has been associated with women: home economics, dealing with the administration and design of family life, or nursing, dealing with the daily care and comfort of patients. Nursing, seen as an extension of the maternal role, has been judged as having no legitimate claim to scientific knowledge. It is important to analyze who determines what counts as science, by what criteria, and within what historical contexts.[15]

These are a few of the analytics that have informed feminist revisions of science. Many of them are standard tools of academic inquiry—and yet the scholarship they have produced has remade disciplines. In my own field of history, for example, the history of women and gender has become an orthodox part of the discipline; a professor who did not employ gender as a category of analysis in a course would be considered irresponsible. Much of this new scholarship has been produced with what we might think of as standard historical methods—scrutiny of archival materials, textual analysis, amassing of indirect indicators for certain demographic trends, and so on. The questions asked, however, have been radically different and have led to the questioning of basic assumptions about what counts as history.

Government Action

Action within the academy relies on receptive audiences and appropriate funding. U.S. funding agencies hold tremendous power to advance equality for women in science. Bernadine Healy, a former head of NIH, put it simply: "Let's face it, the way to get scientists to move into a certain area is to fund that area." As we have seen, advances in research on women's health in the United States were reinforced by laws requiring grant applications to include female participants in medical research (or to justify their exclusion). In 1994 the National Science Foundation (NSF) reduced funding to the Aspen Center for Physics because the Center (it was told informally) was not doing enough to increase its numbers of women; in response the Aspen Center held its first weeklong meeting devoted to women's issues and adopted many of the participants' recommendations (including an increase in the number of women on its scientific advisory board, attention to representation of women in nominations, more day-care options, and efforts to allow couples to coordinate visits). Women's professional visibility was enhanced when Mary Clutter, NSF's assistant director for biological sciences, made it known that conference organizers need not apply for support if their conferences included no women as invited speakers.[16]

Private agencies and individuals also hold the power of the purse. Alumnae of Radcliffe College made news in 1995 when they placed all their contributions to their alma mater in an escrow account to be held until Harvard hired more women faculty members.

Efforts at NSF to foster gender equality in science, however, pale in comparison with those at NIH. In the 1990s NSF began a number of programs for women, including the Program for Women and Girls, the Visiting Professorships for Women, Faculty Awards for Women, Research Planning Grants for Women, and Career Advancement for Women, many of which have now been consolidated into the program called Professional Opportunities for Women in Research and Education. These programs focus on career advancement for women scientists, an issue of crucial importance. But no section at NSF is designated to oversee the removal of gender bias from basic research. At NIH, in contrast, women's career advancement is intimately tied to the correction of bias in research.

Some scientists will object that NSF does "basic" research, different in kind from that pursued by NIH. But gender bias can be as real in basic research as in applied research. When zoologists routinely study gene receptors only in male animals and field biologists routinely release all the females trapped with the animals intended for a particular study, is basic research telling the whole story? There is more to correcting research design than including omitted females. But that is conceptually an easy place to begin. Talking to research scientists, however, makes it clear that funding agencies must also understand the cost of including females. Including females in a study creates a need for more control groups: required are prepubescent and mature males and a control group for each, plus corresponding groups of females and a control group for each.

In the United States there are currently many redundant efforts to improve the situation for women in science. Many universities have special programs for women in science and engineering,[17] but too many are just window dressing. Universities are happy to have a member of the faculty or the staff devoted to recruiting women students, but balk at programs designed to change the internal cultures of the science departments these women will be entering. While some universities support curriculum reform, none have programs devoted to uncovering bias in research. Universities encourage women students and faculty members to donate their time to "mentor" each other and provide support systems, but these efforts tend to be sporadic, dependent on soft money and volunteer efforts. Under such conditions, universities are asking women to remedy the historical shortcomings of the academy by bearing the burden of creating a welcoming environment for themselves.

There is, perhaps, some room for hope. A nationally coordinated effort with some legislative power may be in the offing. The Morella Commission (named for Representative Constance Morella, a Maryland Republican) has called for a full review of women in science; a federal bill proposed in 1993 would set up a commission to study the problems women face in entering and succeeding in technical professions. While no action has yet been taken (two bills are still in committee), the groundwork has been laid. One problem is that the powerful Women's Caucus was formally disbanded by the Republicans in the mid-1990s. Though the Caucus now has no office or budget, it still informally oversees congressional action on a number of issues, ranging from the Beijing Conference on Women to domestic violence, women's health, and women's education.

Overseas there are similar efforts. In 1994 the United Kingdom published a national report on women in science and engineering, *The Rising Tide,* with specific proposals for fostering women's careers in these areas.[18] In 1996 the German federal government held an international meeting on women in the sciences. Lower Saxony's Ministry for Science and Culture recently published a report on gender research in the sciences, engineering, and medicine prepared by leading women scientists and coordinated by the minister, Helga Schuchardt. The report called for the opening of a women's university, modeled on U.S. women's colleges but incorporating feminist research and pedagogy. A model Women's University for Engineering and Cultures is planned to run for three months in Hannover in the year 2000 as part of the World's Fair. And such efforts are now being coordinated across Europe: in the spring of 1998 the European Union set up a new commission to oversee efforts to improve the status of women in European science.[19]

Each of these projects combines research on women and gender with government initiatives. This kind of mission-oriented science is familiar. The Manhattan Project was government-directed science aimed at securing national defense. The Apollo Program to land men on the moon, the attempt to build, launch, and operate a space station, and the costly Human Genome Project, a fifteen-year research effort to map the human genome, are all examples of mission-oriented government-funded science. The U.S. Congress should launch a "women's science and engineering initiative" to support analysis of gender in the content of the sciences and to promote equality for women in scientific and technical fields. This initiative should be a collaborative effort joining the expertise of scientists, anthropologists, historians, and theorists.

Society and Culture

Americans hold individualism sacred. Human beings are, however, not isolated individuals but exist in various webs of professional networks and personal relationships. Historically, the word "individuals" has meant male heads of households, so that our very notion of an individual embodies a division of social and intellectual labor that put men in the workplace and women in the home. This false notion of individualism is more significant professionally for women than for men, because more professional women than professional men are part of dual-career couples.

The past decade has seen a plethora of suggestions for integrating women into professional life: from the professionally debilitating "mommy track" to the belated passage of the inadequate family leave act to the hiring of both members of dual-career couples. All these initiatives are welcome, but they leave many basic structures unchanged. Sexual divisions in physical and intellectual labor structure institutions, technology, and everyday objects. While grocery carts and strollers have been designed to fit women's bodies (people taller than the typical height of women often find pushing them uncomfortable), cockpits and artificial hearts have been designed to fit men's.[20] Public buildings, too, have been designed for men, or at least not for women. Where, for example, are lactating rooms in public institutions? Former Speaker of the House of Representatives Newt Gingrich set aside a room for House mothers to breastfeed, but this is a rare exception.

Though the situation of women has improved tremendously, American and European societies persist in using foundational divisions between domestic and professional life that date from the eighteenth century. It is worth noting that other organizations of social life can benefit working men and women. In 1700, 14 percent of German astronomers were women, a higher percentage than in Germany or the United States today. As we saw in Chapter 1, this was possible because astronomy was a household industry. I am not suggesting that we return to premodern economic structures or that early modern guilds were havens for women. Artisanal women were wifely assistants; and while some enjoyed a large measure of independence, most were subordinate to their husbands. My point is that different forms of organizing working and private lives yield different results for women.

As women enter the professions and men take increasing responsibility in the home, the relationship between professional and private life will be rethought and restructured. Gender differences have been wrought by historical circumstances. No sleight of an invisible hand of the market will make them disappear. Culture is about unspoken rules. Once we articulate those rules we can begin to reform them to meet new expectations and new needs.

There is no easy solution to questions about gender in science. Feminists have no more of an inside track to the truth than anyone else. There is no firm starting point for change—no Archimedean point—which once established will ensure progressive reform, unless it is a critical understanding of the problem. Such an understanding, I have argued, is in large part available. Feminists have tended to make a distinction between getting women into science and changing knowledge. Getting women in is generally considered the easier of the two tasks. Both, however, depend on proper tools of gender analysis. Both are institutional *and* intellectual problems. Bringing feminism into science will require hard-fought battles in a complex process of political and social change. Science departments cannot solve the problems themselves because the problems are also deeply cultural. But that does not let them off the hook. Change will have to happen simultaneously in many areas, including conceptions of knowledge and research priorities, domestic relations, attitudes in preschools and schools, structures at universities, practices in classrooms, the relationship between home life and the professions, and the relationship between our culture and others.

Appendix

Table 1. Territoriality: How women cluster in scientific disciplines

Field of Study	1990	1993	1996	% Women Ph.D.'s 1996
Sciences, total	5,955	7,130	7,870	37.6
Physical science, total	661	780	842	21.9
Astronomy, total	20	25	41	21.4
Astronomy	13	12	21	25.0
Astrophysics	7	13	20	18.5
Chemistry, total	503	582	605	28.2
Analytical	68	91	111	32.1
Inorganic	60	59	68	27.3
Medicinal/pharmaceutical	20	38	32	33.3
Nuclear	2	0	3	0
Organic	104	131	118	23.3
Physical	76	88	85	28.3
Polymer	13	22	30	24.8
Theoretical	14	16	14	24.6
Chemistry, general	128	112	119	30.1
Chemistry, other	18	25	28	38.9
Physics, total	130	169	193	13.0
Acoustics	2	1	2	10.5
Chemical and atomic/ molecular	7	14	10	7.8

(continued)

Table 1. (continued)

Field of Study	1990	1993	1996	% Women Ph.D.'s 1996
Electron	0	n/a	n/a	—
Elementary particle	14	18	19	10.9
Fluids	1	2	4	19.0
Nuclear	5	8	9	10.3
Optics	5	14	20	15.5
Plasma and high-temperature	2	4	2	4.2
Polymer	3	5	12	36.4
Solid state and low-temperature	35	42	54	14.8
Physics, general	33	46	39	12.0
Physics, other	23	15	22	14.1
Other physical sciences	8	4	3	23.1
Earth, atmospheric, and ocean sciences, total	141	160	172	21.7
Atmospheric sciences, total	14	18	22	17.6
Atmospheric dynamics	4	5	4	19.0
Atmospheric physics/ chemistry	1	2	7	31.8
Meteorology	3	3	2	5.7
Atmospheric science/ meteorology, general	5	7	5	15.2
Atmospheric science/ meteorology, other	1	1	4	28.6
Geosciences, total	85	89	88	19.5
Applied geology	3	n/a	n/a	—
Geology	32	35	36	22.2
Geochemistry	9	13	10	20.4
Geomorphology/glacial geology	1	3	1	9.1
Geophysics/seismology	15	14	14	13.9
Hydrology/water resources	1	9	5	16.1
Mineralogy/petrology	4	1	11	47.8
Paleontology	2	5	4	28.6
Stratigraphy/sedimentation	5	5	3	25.0
Geological and related sciences, general	5	2	2	7.4

Table 1.   (continued)

Field of Study	1990	1993	1996	% Women Ph.D.'s 1996
Geological and related sciences, other	8	2	2	9.1
Oceanography, total	31	35	39	29.1
Marine sciences	7	8	5	18.5
Oceanography	24	27	34	31.8
Other environmental sciences	11	18	23	27.7
Mathematics/computer sciences total	268	402	370	18.1
Mathematics, total	158	264	231	20.6
Algebra	6	28	18	23.1
Analysis/function analysis	15	23	15	15.0
Applied mathematics	34	43	52	22.6
Computing theory	1	3	2	11.1
Geometry	8	7	14	19.4
Logic	3	4	1	6.3
Mathematical statistics	31	57	47	26.4
Number theory	1	4	7	16.7
Operations research	2	9	4	19.0
Topology	11	9	5	9.1
Mathematics, general	32	56	45	19.3
Mathematics, other	14	21	21	26.6
Computer sciences, total	110	138	139	15.1
Computer science	85	123	116	13.9
Information science/systems	25	15	23	27.4
Biological sciences, total	1,615	2,050	2,415	42.2
Bacteriology	4	6	8	50.0
Biochemistry	237	318	317	39.9
Biomedical sciences	n/a	n/a	54	38.6
Biophysics	23	21	41	28.9
Biotechnology research	n/a	4	1	16.7
Plant genetics	7	12	16	39.0
Plant pathology	12	15	13	34.2
Plant physiology	19	19	29	39.7
Botany, other	33	43	44	41.9
Anatomy	29	34	20	42.6

(continued)

Table 1. (continued)

Field of Study	1990	1993	1996	% *Women* *Ph.D.'s* *1996*
Biometrics/biostatistics	18	21	34	42.0
Cell biology	69	116	107	45.9
Developmental biology/ embryology	12	25	49	51.0
Ecology	52	67	84	34.3
Endocrinology	11	12	13	54.2
Entomology	26	28	36	26.5
Biological immunology	72	73	129	54.2
Microbiology	116	183	185	41.7
Molecular biology	152	237	291	44.7
Neuroscience	76	110	165	40.8
Nutritional sciences	88	99	98	69.0
Parasitology	4	7	12	54.5
Toxicology	36	43	60	43.5
Human/animal genetics	73	76	101	47.6
Human/animal pathology	32	45	52	38.5
Human/animal pharmacology	92	121	142	44.9
Human/animal physiology	98	100	107	38.9
Zoology, other	37	32	31	31.0
Biological sciences, general	140	113	120	41.2
Biological sciences, other	47	70	56	40.6

Source: National Science Foundation, *Science and Engineering Doctorate Awards: 1996* (Arlington, Va., 1997), 12–14, 26–28, 30.

Table 2. Occupational territoriality: Women by ethnicity

Ethnicity	Field	Number of Ph.D.'s		
		1990	1993	1996
African American	Sciences, total	143	197	234
	Physical sciences, total	7	10	13
	Astronomy	0	0	0
	Chemistry	7	9	10
	Physics	0	0	3
	Other physical sciences	0	1	0
	Earth, atmospheric, and ocean sciences	2	0	2
	Mathematics	1	1	2
	Computer sciences	0	4	3
	Biological sciences	22	32	40
	Psychology	72	75	115
	Social sciences	38	71	59
Asian/Pacific Islander	Sciences, total	219	470	850
	Physical sciences, total	42	97	145
	Astronomy	2	2	4
	Chemistry	29	72	108
	Physics	11	23	33
	Other physical sciences	0	0	0
	Earth, atmospheric, and ocean sciences	1	8	12
	Mathematics	7	36	36
	Computer sciences	7	19	16
	Biological sciences	87	172	408
	Psychology	31	51	93
	Social sciences	38	78	110
European American	Sciences, total	4,296	4,751	5,006
	Physical sciences, total	385	388	431
	Astronomy	16	21	30
	Chemistry	315	299	308
	Physics	48	66	91
	Other physical sciences	6	2	2
	Earth, atmospheric, and ocean sciences	109	110	109

(*continued*)

Table 2. (continued)

Ethnicity	Field	Number of Ph.D.'s 1990	1993	1996
	Mathematics	79	124	105
	Computer sciences	80	83	61
	Biological sciences	1,154	1,311	1,401
	Psychology	1,573	1,711	1,836
	Social sciences	770	909	926
Hispanic	Sciences, total	169	210	258
	Physical sciences, total	26	20	12
	Astronomy	0	0	1
	Chemistry	24	18	7
	Physics	2	2	4
	Other physical sciences	0	0	0
	Earth, atmospheric, and ocean sciences	4	2	6
	Mathematics	4	2	0
	Computer sciences	1	1	7
	Biological sciences	33	57	66
	Psychology	57	86	120
	Social sciences	38	37	43
Native American	Sciences, total	18	18	32
	Physical sciences, total	0	2	1
	Astronomy	0	0	0
	Chemistry	0	1	1
	Physics	0	1	0
	Other physical sciences	0	0	0
	Earth, atmospheric, and ocean sciences	1	0	0
	Mathematics	0	0	0
	Computer sciences	0	1	2
	Biological sciences	1	2	9
	Psychology	13	9	9
	Social sciences	3	4	8

Source: National Science Foundation, *Science and Engineering Doctorate Awards: 1996* (Arlington, Va., 1997), 12–14, 26–28, 30. Includes U.S. citizens and permanent residents only.

Notes

Introduction

1. Curie is the only woman so honored for her own merits. Sophie Berthelot, the other woman interred in the Pantheon, lies there with her husband, a renowned French chemist, who died of grief one hour after her death.

2. Gross and Levitt, *Higher Superstition,* 110; Paul Gross, Norman Levitt, and Martin Lewis, eds., *The Flight from Science and Reason* (New York: New York Academy of Sciences, 1996); Noretta Koertge, "Are Feminists Alienating Women from the Sciences?" *Chronicle of Higher Education* (14 Sept. 1994): A80. Briscoe, "Scientific Sexism," 153.

3. Others have discussed the intricacies of feminist theory and its relation to science; see Sue Rosser, "Possible Implication of Feminist Theories for the Study of Evolution," in *Feminism and Evolutionary Biology,* ed. Gowaty; Longino, "Subjects"; Harding, *Science Question.*

4. John Barry and Evan Thomas, "Military: At War over Women," *Newsweek* (12 May 1997).

5. Harding, *Science Question,* 24–25.

6. Oelsner, *Die Leistungen,* 3–5. Bruno Bettelheim, "The Commitment Required of a Woman Entering a Scientific Profession in Present-Day American Society," in *Women and the Scientific Professions,* ed. Jacquelyn Mattfeld and Carol Van Aken (Cambridge, Mass.: MIT Press, 1965), 18.

7. Mary Belenky, Blythe Clinchy, Nancy Goldberger, and Jill Tarule, *Women's Ways of Knowing: The Development of Self, Voice, and Mind* (New York: Basic Books, 1986); Nancy Goldberger, Jill Tarule, Blythe Clinchy, and Mary Belenky, eds., *Knowledge, Difference, and Power: Essays Inspired by Women's Ways of Knowing* (New York: Basic Books, 1996); Nel Noddings, *Caring: A Feminine Approach to Ethics and Moral Education* (Berkeley: University of California Press, 1984); Rose, "Hand, Brain, and Heart;" Sara Ruddick, *Maternal Thinking: Toward a Politics of Peace* (Boston: Beacon, 1989);

Carol Gilligan, *In a Different Voice: Psychological Theory and Women's Development* (Cambridge, Mass.: Harvard University Press, 1982).

8. Judith Butler, *Gender Trouble: Feminism and the Subversion of Identity* (New York: Routledge, 1990); Haraway, *Simians*.

9. N. Katherine Hayles, *Chaos Bound: Orderly Disorder in Contemporary Literature and Science* (Ithaca: Cornell University Press, 1990).

10. Many people make this point: e.g., Haraway, *Simians*; Longino, "Cognitive and Non-Cognitive Values," 49.

11. Rowell, "Introduction," 16; see Hrdy, "Empathy," 134–139.

12. Evelyn Fox Keller, *A Feeling for the Organism: The Life and Work of Barbara McClintock* (San Francisco: Freeman, 1983), 198. Keller, *Reflections*, 158–179.

13. Keller, *Secrets*, 32–33. Fedigan and Fedigan, "Gender and the Study of Primates," 45.

14. Haraway, *Primate Visions*. Hrdy, "Empathy," 137.

15. Stephen Jay Gould, "The Triumph of a Naturalist," *New York Review of Books* (29 March 1984).

16. Jeanne Altmann, "Observational Study of Behavior: Sampling Methods," *Behaviour* 49 (1974).

17. Helen Longino, "Can There Be a Feminist Science?" in *Feminism and Science*, ed. Tuana; Conkey, "Making the Connections," 4.

18. Barinaga, "Female Style"; John Benditt, "Editor's Note," *Science* 261 (23 July 1993).

19. While most Americans agree that the women's movement has improved the status of women, the number of women who consider "feminist" an insult has increased since 1992, while the number who think it a compliment has been cut in half. CBS News Poll, 18–20 Sept. 1997.

20. Mildred Dresselhaus, "Women Graduate Students," *Physics Today* 39 (June 1986); Rosabeth Kanter, *Men and Women of the Corporation* (New York: Basic Books, 1977).

21. Sonnert and Holton used questionnaire data from 191 women (all white) and 508 men (98 percent white) and interviews with 108 women and 92 men. Sonnert and Holton, *Gender Differences*, 33–34, 142–155. Most reported no evidence of a distinctively "female methodology or way of thinking"; women may employ standard methodologies more cautiously or meticulously, but they do not employ a radically different, nonandrocentric methodology. The authors caution that these findings are based on scientists' perceptions and self-reports.

22. Donna Holmes and Christine Hitchcock, "A Feeling for the Organism? An Empirical Look at Gender and Research Choices of Animal Behaviorists," in *Feminism and Evolutionary Biology*, ed. Gowaty. Holmes and Hitchcock (196–197) cite Ted Burk, who surveyed articles published in *Animal Behaviour* between 1953 and 1993 and found women more likely than men to study mammals (including primates), sexual selection, mate choice, infants or juveniles, and maternal care.

23. Haraway, *Primate Visions*, 316.

24. Stamps, "Role of Females," 294.
25. Barinaga, "Surprises," 14468. Committee on Women in Science and Engineering, *Women Scientists and Engineers*, 32.
26. William Whewell, "On the Connexion of the Physical Sciences, by Mrs. Somerville," *Quarterly Review* 51 (March 1834): 65. See Robert Merton, "De-Gendering 'Man of Science': The Genesis and Epicene Character of the Word *Scientist*," in *Sociological Visions*, ed. Kai Erikson (Lanham, Md.: Rowman and Littlefield, 1997).
27. Snow, *Two Cultures*, 4. Joan Landes, *Women and the Public Sphere in the Age of the French Revolution* (Ithaca: Cornell University Press, 1988); Christine Fauré, *Democracy without Women: Feminism and the Rise of Liberal Individualism in France*, trans. Claudia Gorbman and John Berks (Bloomington: Indiana University Press, 1991).
28. Stephen Weinberg, "Reflections of a Working Scientist," *Daedalus* (Summer 1974): 43.
29. Krieger and Zierler, "Accounting for Health of Women," 253.
30. Keller, *Secrets*, 33.

1. Hypatia's Heritage

1. Cited in Wilfrid Blunt, *The Compleat Naturalist: A Life of Linnaeus* (London: William Collins Sons, 1971), 157.
2. Pizan, *Book of the City of Ladies*, 70–71.
3. Giovanni Boccaccio, *De mulieribus claris*, trans. Guido Guarino as *Concerning Famous Women* (New Brunswick: Rutgers University Press, 1963). Some encyclopedias: Augustin della Chiesa, *Theatrum literatar feminarum* (1620); Johann Frauenlob, *Die Lobwürdige Gesellschaft der gelehrten Weiber* (1631); Marguerite Buffet, *Eloges des illustres sçavantes anciennes et modernes* (1668); J. C. Eberti, *Eröffnetes Cabinet des gelehrten Frauenzimmers* (1706); C. F. Paullini, *Hoch- und Wohlgelahrtes teutsches Frauenzimmer* (1712); Gilles Ménage, *Historia mulierum philosopharum* (1690), trans. Beatrice Zedler as *The History of Women Philosophers* (Lanham, Md.: University Press of America, 1984).
4. Jérôme de Lalande, *Astronomie des dames* (1786; Paris, 1820), 5–6. Christian Harless, *Die Verdienste der Frauen um Naturwissenschaft, Gesundheits- and Heilkunde* (Göttingen, 1830), ix and 2.
5. Darwin, *Descent*, vol. 2, 327. Gino Loria, "Les Femmes mathématiciennes," *Revue scientifique* 20 (1903): 386.
6. H. J. Mozans, *Woman in Science* (1913; Cambridge, Mass.: MIT Press, 1974), 391, 415–416.
7. A. W. Richeson, "Hypatia of Alexandria," *Natural Mathematics Magazine* 15 (1940); Marie-Louise Dubreil-Jacotin, "Figures de mathématiciennes," in *Les Grands Courants de la pensée mathématique*, ed. F. le Lionnais (Marseille: Cahiers du Sud, 1948); Julian Coolidge, "Six Female Mathematicians," *Scripta Mathematica* 17 (1951); Denis Duveen, "Madame Lavoisier: 1758–1836," *Chymia Annual: Studies in the History of Chemistry* 4 (1953);

V. Rizzo, "Early Daughters of Urania," *Sky and Telescope* 14 (1954); Edna Yost, *Women of Modern Science* (New York: Dodd, 1959).

8. Marie Curie, "Autobiographical Notes," in *Pierre Curie,* trans. Charlotte and Vernon Kellogg (New York: Macmillan, 1923); Ida Hyde, "Before Women Were Human Beings: Adventures of an American Fellow in German Universities of the '90s," *Journal of the American Association of University Women* 31 (1938); Lise Meitner, "The Status of Women in the Professions," *Physics Today* 13 (1960); Kathleen Lonsdale, "Women in Science: Reminiscences and Reflections," *Impact of Science on Society* 20 (1970); Vivian Gornick, *Women in Science: Portraits from a World in Transition* (New York: Simon and Schuster, 1983); Derek Richter, ed., *Women Scientists: The Road to Liberation* (London: Macmillan, 1982); Naomi Weisstein, "Adventures of a Woman in Science," in *Biological Woman,* ed. Hubbard, Henifin, and Field; *Cecilia Payne-Gaposchkin,* ed. Haramundanis; Ajzenberg-Selove, *A Matter of Choices;* Susan Ambrose, Kristin Dunkle, Barbara Lazarus, Indira Nair, Deborah Harkus, *Journeys of Women in Science and Engineering: No Universal Constants* (Philadelphia: Temple University Press, 1997).

9. E.g., Robert Reid, *Marie Curie* (New York: Saturday Review Press, 1974); Anne Sayre, *Rosalind Franklin and DNA* (New York: Norton, 1975); Olga Opfell, *The Lady Laureates: Women Who Have Won the Nobel Prize* (Metuchen, N.J.: Scarecrow, 1978); Louis Bucciarelli and Nancy Dworsky, *Sophie Germain: An Essay in the History of the Theory of Elasticity* (Dordrecht: Reidel, 1980); James Brewer and Martha Smith, eds., *Emmy Noether: A Tribute to Her Life and Work* (New York: Dekker, 1981); Elizabeth Patterson, *Mary Somerville and the Cultivation of Science, 1815–1840* (The Hague: Nijhoff, 1983); Ann Hibner Koblitz, *A Convergence of Lives, Sofia Kovalevskia: Scientist, Writer, Revolutionary* (Boston: Birkhäuser Boston, 1983); Alic, *Hypatia's Heritage;* Kass-Simon and Farnes, eds., *Women of Science;* McGrayne, *Nobel Prize Women;* Cheryl Claassen, ed., *Women in Archaeology* (Philadelphia: University of Pennsylvania Press, 1994); Maria Dzielska, *Hypathia of Alexandria,* trans. F. Lyra (Cambridge, Mass.: Harvard University Press, 1995); Susan Quinn, *Marie Curie: A Life* (New York: Simon and Schuster, 1995); Theresa Gómez and Gloria Conde, eds., *Mujeres de Ciencia: Mujer, Feminismo y Ciencias Naturales, Experimantales y Technólogias* (Granada: Universidad de Granada, 1996); Benjamin and Barbara Shearer, eds., *Notable Women in the Life Sciences: A Biographical Dictionary* (Westport, Conn.: Greenwood, 1996); Sime, *Lise Meitner;* Joy Harvey, *"Almost a Man of Genius":* Clémence Royer, Feminism, and Nineteenth-Century Science (New Brunswick: Rutgers University Press, 1997).

10. Charles Burney, *The Present State of Music in France and Italy* (1773), ed. Percy Scholes (New York: Oxford University Press, 1959), 159.

11. Benedict to Agnesi, Sept. 1750, cited in Rebière, *Les Femmes,* 11. Edna Kramer, "Maria Gaetana Agnesi," in *Dictionary of Scientific Biography,* ed. Charles Gillispie (New York: Scribner, 1970); Lynn Osen, *Women in Mathematics* (Cambridge, Mass.: MIT Press, 1974). Morandi was employed by the university to dissect and prepare bodies in order to teach anatomy to students

and curious amateurs. Marta Cavazza, "Dottrici' e Lettrici dell'Università de Bologna nel settecento," *Annali di Storia delle Università Italiane* 1 (1997): 120. Maria Dalle Donne held the post of director of the Scuola per Levatrici (School of Midwives) from 1804 to 1842 and was for many years a member of the Istituto delle Scienze.

12. Paula Findlen, "Science as a Career in Enlightenment Italy: The Strategies of Laura Bassi," *Isis* 84 (1993): 449; Beate Ceranski, *"Und Sie Fürchtet sich vor Niemandem": Die Physikerin Laura Bassi, 1711–1778* (Frankfurt: Campus, 1996).

13. Henry Curzon, *The Universal Library: or, Compleat Summary of Science* (London, 1712), vol. 1, 439. Kathleen Lonsdale and Marjory Stephenson were elected to the Royal Society in 1945. Joan Mason, "The Admission of the First Women to the Royal Society of London," *Notes and Records of the Royal Society of London* 46 (1992).

14. Pierre Remy, *Catalogue d'une collection de très belles coquilles, madrepores, stalactiques, . . . de Madame Bure* (Paris, 1763); Jacques Roger, *Les Sciences de la vie dans la pensée française du XVIIIᵉ siècle* (Paris: Armand Colin, 1963). Science for ladies remained popular throughout Europe in the eighteenth century. In Italy, the poet Francesco Algarotti published an introduction to Newtonian physics in 1737. In Germany, Johanna Charlotte Unzer published her *Grundriss einer Weltweisheit für Frauenzimmer* in 1761; in Russia, Leonhard Euler wrote his *Letters to a German Princess on Diverse Points of Physics and Philosophy* in 1768; Gerald Meyer, *The Scientific Lady in England: 1650–1760* (Berkeley: University of California Press, 1955).

15. René Taton, "Gabrielle-Émilie le Tonnelier de Breteuil, Marquise du Châtelet," in *Dictionary of Scientific Biography;* Elizabeth Badinter, *Emilie, Emilie: L'Ambition féminine au XVIIIᵉ siècle* (Paris, 1983); Linda Gardiner, "Women in Science," in *French Women and the Age of Enlightenment,* ed. Samia Spencer (Bloomington: Indiana University Press, 1984); Mary Terrall, "Emilie du Châtelet and the Gendering of Science," *History of Science* 33 (1995).

16. *Carpenrariana or remarques . . . de M. Charpentier* (Paris, 1724), 316; Claude Clerselier, *Lettres de Mr. Descartes* (Paris, 1724), vol. 1, preface.

17. Lougee, *Le Paradis des femmes,* 41–53; Dena Goodman, "Enlightenment Salons: The Convergence of Feminine and Philosophical Ambitions," *Eighteenth-Century Studies* 22 (1989); Schiebinger, *Mind,* 30–32; Findlen, "Translating the New Science."

18. Pizan, *Book of the City of Ladies,* 70–80. Edgar Zilsel, "The Sociological Roots of Science," *American Journal of Sociology* 47 (1942).

19. Schiebinger, *Mind,* ch. 3. On Merian see also Natalie Zemon Davis, *Women on the Margins: Three Seventeenth-Century Lives* (Cambridge, Mass.: Harvard University Press, 1995).

20. Donnison, *Midwives;* Marland, ed., *Art of Midwifery.* Alic, *Hypatia's Heritage,* 88–92.

21. See Lawrence Stone, *The Family, Sex, and Marriage in England, 1500–1800* (New York: Harper and Row, 1977); Jean-Louis Flandrin, *Families in Former*

Times: Kinship, Household, and Sexuality, trans. Richard Southern (1975; Cambridge: Cambridge University Press, 1979).

22. Abir-Am and Outram, eds., *Uneasy Careers and Intimate Lives,* intro.

23. Helena Pycior, Nancy Slack, and Pnina Abir-Am, eds., *Creative Couples in the Sciences* (New Brunswick: Rutgers University Press, 1996); Ann Shteir, *Cultivating Women, Cultivating Science: Flora's Daughters and Botany in England, 1760–1860* (Baltimore: Johns Hopkins University Press, 1996).

24. Hynes, "Toward a Laboratory of One's Own."

25. Rossiter, *Women Scientists* (1982), ch. 3; Galison, "Fortran," 228–229.

26. Though some institutions admitted women to graduate work as early as 1877, the more prestigious institutions were slow in doing so. Roy MacLeod and Russell Moseley, "Fathers and Daughters: Reflections on Women, Science and Victorian Cambridge," *History of Education* 8 (1979). Rossiter, *Women Scientists* (1982), 131–132; LaFollette, *Making Science,* 82; Zuckerman et al., eds., *Outer Circle,* 12–13. Mary Roth Walsh, *Doctors Wanted: No Women Need Apply* (New Haven: Yale University Press, 1977).

27. Rossiter, *Women Scientists* (1995), 36.

28. Ibid., 31, 34.

29. *Climbing the Academic Ladder,* 135–136. Briscoe, "Scientific Sexism," 153. Davis and Rosser, "Program and Curricular Interventions." National Science Foundation [henceforth NSF], *Characteristics of Doctoral Scientists and Engineers,* 30.

2. Meters of Equity

1. Daryl Chubin and Shirley Malcom, "Policies to Promote Women in Science," in *Equity Equation,* ed. Davis et al., 7; Paula Rayman and Jennifer Jackson, "Women Scientists in Industry," ibid.; Committee on Women in Science and Engineering, *Women Scientists and Engineers.*

2. Rossiter, *Women Scientists* (1982), ch. 10. Vetter, *Professional Women,* 172. NSF, *Science and Engineering Doctorate Awards: 1996* (Arlington, Va., 1997), 12, 16 (NSF includes psychology, economics, political science, and sociology in its definition of science). Mary Cage, "Women Say opportunities in Engineering Are Improving, but the Pace Is Slow," *Chronicle of Higher Education* (7 April 1995): A20.

3. Rossiter, *Women Scientists* (1982), ch. 8; Rossiter, *Women Scientists* (1995), table 4.4. American Association of University Women, *How Schools Shortchange Girls,* 4.

4. National Center for Education Statistics, *Digest of Education Statistics* (Washington: U.S. Department of Education, 1996), 258–264. Rossiter, *Women Scientists* (1982), 134–137. NSF, *Women, Minorities* (1996), 63.

5. Patricia Ostertag and J. Regis McNamara, "'Feminization' of Psychology: The Changing Sex Ratio and Its Implications for the Profession," *Psychology of Women Quarterly* 15 (1991); Judith Lorber, "A Welcome to a Crowded Field: Where Will the New Women Physicians Fit In?" *Journal of the American*

Medical Women's Association 42 (1987); Constance Holden, "Researchers Find Feminization a Two-Edged Sword," *Science* 271 (29 March 1996).

6. Vetter, *Professional Women,* 251. From the mid 1970s to the mid-1980s men received 0–6 percent of nursing Ph.D.'s and women 2–6 percent of engineering Ph.D.'s. In 1992 women earned 19 percent of the Ph.D.'s in the physical sciences but held only 3 percent of full professorships, earned 39 percent of the Ph.D.'s in the life sciences but held only 10 percent of the full professorships, and earned 48 percent of the Ph.D.'s in the social sciences but held only 11 percent of the full professorships. Florence Denmark, "Engendering Psychology," *American Psychologist* 49 (1994).

7. Zuckerman, "Careers," 39; NSF, *Women, Minorities* (1996), 70.

8. NSF, *Women, Minorities* (1996), 72–74. Edward Silverman, "New NSF Report on Salaries of Ph.D.'s Reveals Gender Gaps in All Categories," *Scientist* 5 (19 Aug. 1991): 20. Edward Silverman, "NSF's Ph.D. Salary Survey Finds Minorities Earn Less than Whites," *Scientist* 5 (16 Sept. 1991): 21.

9. Jeanhee Kim, "Female Engineers: Short Circuit in Pay," *Working Woman* (Dec. 1993): 16.

10. NSF, *Women, Minorities* (1996), 73–74. Barbara Mandula, "Women Scientists Still Behind," *Association for Women in Science Magazine* 20 (May/June 1991): 10–11.

11. Families and Work Institute, *Women, the New Providers: A Study of Women's Views on Family, Work, Society and the Future* (New York, 1995). Diane Harris, "How Does Your Pay Stack Up?" *Working Woman* (Feb. 1996): 27–28. Men's wages dropped by 12 percent between 1973 and 1993, while women's wages rose by 6 percent.

12. NSF, *Women, Minorities* (1996) 75, 106, 108. Ronald Hoy, "A 'Model Minority' Speaks Out on Cultural Shyness," *Science* 262 (12 Nov. 1993): 1117–18.

13. Hull et al., eds., *All the Women Are White.* NSF, *Women and Minorities* (1990), 82. See also Beatriz Clewell and Angela Ginorio, "Examining Women's Progress in the Sciences from the Perspective of Diversity," in *Equity Equation,* ed. Davis et al.; Daniel Solorzano, "The Baccalaureate Origins of Chicana and Chicano Doctorates in the Physical, Life, and Engineering Sciences: 1980–1990," *Journal of Women and Minorities in Science and Engineering* 1 (1994); Beatriz Clewell and Bernice Anderson, *Women of Color in Mathematics, Science, and Engineering* (Washington: Center for Women Policy Studies, 1991).

14. Yitchak Haberfeld and Yehouda Shenhav, "Are Women and Blacks Closing the Gap? Salary Discrimination in American Science during the 1970s and 1980s," *Industrial and Labor Relations Review* 44 (1990). Silverman, "NSF's Ph.D. Salary Survey."

15. Shirley Malcom, "Equity and Excellence: Compatible Goals" (Washington: American Association for the Advancement of Science, 1983). Vivienne Malone Mayes, "Black and Female," *Association for Women in Mathematics Newsletter* 5 (1975). Kenschaft and Keith, eds., *Winning Women into Mathematics,* 39.

16. "Comparisons across Culture," *Science* 263 (11 March 1994); Motoko Ku-
wahara, "The Participation of Japanese Women in S&T," Research Institute
for Education, St. Andrews University, Japan, March 1998. Barinaga, "Surprises."

17. I thank Annette Vogt, Max-Planck-Institut für Wissenschaftsgeschichte, for
this information. In 1997 14 percent of the scientists at the 76 Max Planck
Institutes were women; only 2 percent of scientific members were women.

18. Regional Development Plan 1994–1999, submitted to the European Union
Social funds, cited by Mary Osborn, "Status and Prospects of Women in
Science in Europe," *Science* 263 (11 March 1994).

19. Faruqui et al., eds., *Role of Women;* Kotte, *Gender Differences in Science.*

20. Xie Xide, "Women Scientists in China: Past, Present and the Future," in *Role
of Women,* ed. Faruqui et al..

21. Feride Acar, "Women in Academic Science Careers in Turkey," in *Women in
Science: Token Women or Gender Equality?,* ed. Veronica Stolte-Heiskanen
(Oxford: Berg, 1991); Patricia Kahn, "Turkey: A Prominent Role on a Stage
Set by History," *Science* 263 (11 March 1994).

22. Chandra Mohanty, "Under Western Eyes: Feminist Scholarship and Colonial
Discourses," in *Third World Women and the Politics of Feminism,* ed. Chandra Mohanty, Ann Rosso, and Lourdes Torres (Bloomington: Indiana University Press, 1991).

23. Helen Appleton, Maria Fernandez, Catherine Hill, and Consuelo Quiroz,
"Gender at the Interface of Science and Technology, and Indigenous Knowledge," Issue Paper for the United Nations Commission on Science and Technology for Development, Gender Working Group, 10 May 1994.

24. Agrawal, "Indigenous and Scientific Knowledge," 3–6. Achoka Awori, "Indigenous Knowledge: Myth or Reality?" *Resources: Journal of Sustainable
Development in Africa* 2 (1991): 1. Sandra Harding, "Is Science Multicultural?" *Configurations* 2 (1994): 319; Sandra Harding, *Is Science Multi-
Cultural?* (Bloomington: Indiana University Press, 1998).

25. Kihika Kiambi and Monica Opole, "Promoting Traditional Trees and Food
Plants in Kenya," in *Growing Diversity,* ed. Cooper, Vellvé, and Hobbelink;
Monica Opole, "Revalidating Women's Knowledge on Indigenous Vegeta-
bles: Implications for Policy," in *Cultivating Knowledge: Genetic Diversity,
Farmer Experimentation, and Crop Research,* ed. Walter de Beof et al. (Lon-
don: Intermediate Technology Publications, 1993). Ram Mahalingam,
"'Feminist Mathematics': Implications for a Multicultural Mathematics Edu-
cation," Women, Gender, and Science Question Conference, University of
Minnesota, Minneapolis, May 1995.

26. Carolyn Sachs, *Gendered Fields: Rural Women, Agriculture, and Environ-
ment* (Boulder: Westview, 1996). Vandana Shiva and Irene Dankelman,
"Women and Biological Diversity: Lessons from the Indian Himalaya," in
Growing Diversity, ed. Cooper et al. Shiva, *Staying Alive,* 65–66.

27. I thank Hector Flores and Carolyn Sachs, Pennsylvania State University, for
calling this example to my attention. See Flores, "Insane Roots and Forked
Radishes: Underground Metabolism, Biotechnology, and Biodiversity,"in

Phytochemicals and Health, ed. David Gustine and Hector Flores (Rockville, Md.: American Society of Plant Physiologists, 1995), 231. One might reflect on the importance of the potato, imported from this area, for the industrial development in the West; the potato was a vital foodstuff for Europe's swelling population in the seventeenth and eighteenth centuries. Alfred W. Crosby, *The Columbian Exchange: Biological and Cultural Consequences of 1492* (Westport, Conn.: Greenwood, 1972), 171.

28. Stephen Brush, "Potato Taxonomies in Andean Agriculture," in *Indigenous Knowledge Systems and Development,* ed. Brokensha et al.; Mario Tapia and Ana de la Torre, *La Mujer Campesina y las Semillas Andinas* (Lima: FAO, 1993).

29. Critics fault the impulse to preserve indigenous knowledge *ex situ* in centralized locations as creating "mausoleums" for knowledge. Agrawal, "Indigenous and Scientific Knowledge," 5. Vandana Shiva, "The Seed and the Earth: Biotechnology and the Colonisation of Regeneration," in *Close to Home: Women Reconnect Ecology, Health and Development Worldwide,* ed. Vandana Shiva (Philadelphia: New Society Publishers, 1994).

30. Bina Agarwal refutes ecofeminism in "The Gender and Environment Debate."

31. Estrella Laredo, "The Advantages and Difficulties of Being a Woman Scientist in a Third World Country," and Gioconda San-Blas, "Venezuelan Women of Science," in *The Role of Women,* ed. Faruqui et al., 726, 739.

32. *Climbing the Academic Ladder,* 19.

33. Cole, *Fair Science,* 69. For opposing views, see Paul Atkinson and Sara Delamont, "Professions and Powerlessness: Female Marginality in the Learned Occupations," *Sociological Review* 38 (1990); Yehouda Shenhav and Yitchak Haberfeld, "Scientists in Organizations: Discrimination Processes in an Internal Labor Market," *Sociological Quarterly* 29 (1988).

34. Cole and Zuckerman, "Productivity Puzzle," 225.

35. Long, "Measures of Sex Differences." Cole and Zuckerman, "Productivity Puzzle," 245, 249.

36. Cole and Zuckerman, "Productivity Puzzle," 218. Mary Frank Fox, "Gender, Environmental Milieu, and Productivity in Science," in *The Outer Circle,* ed. Zuckerman et al., 198.

37. Behrensmeyer quoted in *Science* 255 (13 March 1992): 1388.

38. Cole and Zuckerman, "Marriage," 160.

39. J. Scott Long, "The Origins of Sex Differences in Science," *Social Forces* 68 (1990).

40. J. Scott Long, "Productivity and Academic Position in the Scientific Career," *American Sociological Review* 43 (1978). Nigel Williams, "EU Moves to Decrease the Gender Gap," *Science* 280 (1998): 822. Zuckerman "Careers," 46; Sonnert and Holton, "Glass Ceiling," 6.

41. Robert Merton initially dubbed this phenomenon the "Matthew Effect," referring to the Gospel According to Matthew: "for unto every one that hath shall be given . . . but from him that hath not shall be taken away.": "The Matthew Effect in Science," *Science* 159 (5 Jan. 1968). This notion has been secularized as "cumulative advantage" and "cumulative disadvantage." Mar-

garet Rossiter has created the "Matilda Effect" to describe the special position of women in science: "The (Matthew) Matilda Effect in Science," *Social Studies of Science* 23 (1993).

42. David Hamilton, "Publishing by—and for?—the Numbers," *Science* 250 (7 Dec. 1990): 1331.

43. Cole and Zuckerman, "Productivity Puzzle," 235. Long, "Measures of Sex Differences," 173. Long studied women biochemists for the years 1950–1963. Sonnert and Holton reconfirmed this finding for biologists: women's papers were cited 24.4 times, men's papers 14.1 times: *Gender Differences*, 149. E. Garfield found similar patterns in his study of the thousand most-cited scientists: "Women in Science," *Current Comments* 9 (1 March 1993); as did Cole and Zuckerman: "Productivity Puzzle," 218.

44. Sonnert and Holton, *Gender Differences*, 149–151.

45. Gerbi quoted in Elizabeth Culotta, "Study: Male Scientists Publish More, Women Cited More," *Scientist* (26 July 1993): 14. Sonnert and Holton, *Gender Differences*, 147.

46. Michele Paludi and Lisa Strayer, "What's in an Author's Name? Differential Evaluations of Performance as a Function of Author's Name," *Sex Roles* 12 (1985); Michele Paludi and William Bauer, "Goldberg Revisited: What's in an Author's Name," *Sex Roles* 9 (1993).

47. Storer, "Hard Sciences," 79.

48. Wesley Shrum and Yehouda Shenhav, "Science and Technology in Less Developed Countries," in *Handbook of Science and Technology Studies*, ed. Jasanoff et al.

49. Spector, "Women Astronomers," 20; only 8 percent of minority members said they had witnessed or experienced discrimination against minorities. Vetter, "Glass Ceiling," 13. Spector, "Women Astronomers," 20. Joan Burrelli, "Women Chemists in the US," *Chemistry and Industry* (21 June 1993): 464. Susan Phillips and Margaret Schneider, "Sexual Harassment of Female Doctors by Patients," *New England Journal of Medicine* (23 Dec. 1993). Harassment persists at all educational levels. In a 1987 survey of Harvard faculty and students, 2 percent of the tenured women professors and 49 percent of untenured women reported having been sexually harassed, 1 percent of women graduate students and 34 percent of women undergraduates had been sexually harassed by a person in authority at least once during their time at Harvard, and 15 percent of graduate students and 12 percent of undergraduates said they had changed their majors because of harassment. Most of these faculty members and students said they had not reported the incidents for fear of repercussions. Hewitt and Seymour, "Factors," 98.

50. Goldman-Rakic quoted in Barinaga, "Profile," 1367. Maxon interview with the author aired on WPSU Radio, 1992. *Science* 255 (13 March 1992): 1369.

51. *Sunday World-Herald* (16 June 1991): 13-B. Jane Gross, "Stanford Medical School Official Is Ousted after Sexism Complaint," *New York Times* (25 Feb. 1992): A13. Frances Conley, *Walking Out on the Boys* (New York: Farrar, Straus and Giroux, 1998).

52. Beverly Sauer, "Introduction: Gender and Technical Communication," *IEEE Transactions on Professional Communication* 35 (Dec. 1992): 193–194.
53. Zappert and Stanbury, "Pipeline," 21. Akilah Monifa, "Of African Descent: A Three-fers Story," in *Lesbians in Academia: Degrees of Freedom*, ed. Beth Mintz and Ester Rothblum (New York: Routledge, 1997).
54. Bernice Sandler and Roberta Hall, "The Campus Climate Revisited: Chilly for Women Faculty, Administrators, and Graduate Students" (Washington: Association of American Colleges, 1986), 2.

3. The Pipeline

1. Betty Vetter, "The Science and Engineering Talent Pool," in *Proceedings of the 1984 Joint Meeting of the Scientific Manpower Commission and the Engineering Manpower Commission* (Washington: National Academy of Sciences, May 1984), 2–3.
2. Marilyn Stern and Katherine Karraker, "Sex Stereotyping of Infants: A Review of Gender Labeling Studies," *Sex Roles* 20 (1989). Spertus, *Female Computer Scientists*, 3. Andrée Pomerleau, Daniel Bolduc, Gérard Malcuit, and Louis Cossette, "Pink or Blue: Environmental Gender Stereotypes in the First Two Years of Life," *Sex Roles* 22 (1990).
3. Spertus, *Female Computer Scientists*, 3, 4. Gary Cross, *Kids' Stuff: Toys and the Changing World of American Childhood* (Cambridge, Mass.: Harvard University Press, 1997).
4. Alison Kelly and Barbara Smail, "Sex Stereotypes and Attitudes to Science among Eleven-Year-Old Children," *British Journal of Educational Psychology* 56 (1986): 163. Peter Crabb and Dawn Bielawski, "The Social Representation of Material Culture and Gender in Children's Books," *Sex Roles* 30 (1994).
5. Matyas and Malcom, eds., *Investing in Human Potential*, 20; Jean Grambs and John Carr, *Sex Differences and Learning: An Annotated Bibliography of Educational Research, 1979–1989* (New York: Garland, 1991); Myra and David Sadker, *Failing at Fairness: How America's Schools Cheat Girls* (New York: Scribner, 1994); Sandra Hanson, *Lost Talent: Women in the Sciences* (Philadelphia: Temple University Press, 1996).
6. Joanne Becker, "Differential Treatment of Females and Males in Mathematics Classes," *Journal for Research in Mathematics Education* 12 (1981): 48; Gilah Leder, "Gender and Classroom Practice," in *Gender and Mathematics: An International Perspective*, ed. Leone Burton (London: Cassell, 1990); Susan Gabriel and Isaiah Smithson, *Gender in the Classroom: Power and Pedagogy* (Urbana: University of Illinois Press, 1990), 2–3. Mary Koehler, "Classrooms, Teachers, and Gender Differences in Mathematics," in *Mathematics and Gender*, ed. Fennema and Leder.
7. Dix, ed., *Women*, 23, 110; Linda Grant, "Black Females' 'Place' in Desegregated Classrooms," *Sociology of Education* 57 (1984); Vetter, "Glass Ceiling," 9.
8. Yasmin Kafai, "Electronic Playworlds: Gender Differences in Children's Con-

struction of Video Games," in *Interacting with Video*, ed. Patricia Greenfield and Rodney Cocking (Norwood, N.J.: Ablex, 1996); Charles Huff and Joel Cooper, "Sex Bias in Educational Software: The Effect of Designers' Stereotypes on the Software They Design," *Journal of Applied Social Psychology* 17 (1987); Joel Cooper, Joan Hall, and Charles Huff, "Situational Stress as a Consequence of Sex-Stereotyped Software," *Personality and Social Psychology Bulletin* 16 (1990); Ruth Perry and Lisa Greber, "Women and Computers: An Introduction," *Signs* 16 (1990).

9. Lucy Sells, "High School Mathematics as the Critical Filter in the Job Market," in *Developing Opportunities for Minorities in Graduate Education*, ed. R. T. Thomas (Berkeley: University of California Press, 1973).

10. American Association of University Women, *Shortchanging Girls, Shortchanging America: A Call to Action* (Washington: AAUW Educational Foundation, 1991), 10; Matyas and Malcom, eds., *Investing in Human Potential*, 20. K. Arnold's study cited in A. Pearl, M. Pollack, E. Riskin, B. Thomas, E. Wolf, and A. Wu, "Becoming a Computer Scientist," *Communications of the ACM* 33 (1990): 50.

11. Spertus, *Female Computer Scientists*, 17. V. Crandall, "Sex Differences in Expectancy of Intellectual and Academic Reinforcement," in *Achievement-Related Behaviors in Children*, ed. C. Smith (New York: Russell Sage Foundation, 1969); Sumru Erkut, "Exploring Sex Differences in Expectancy, Attribution, and Academic Achievement," *Sex Roles* 9 (1983); Alexander Astin and Helen Astin, *Undergraduate Science Education: The Impact of Different College Environments on the Educational Pipeline in the Sciences* (Los Angeles: University of California, Higher Education Research Institute, 1993).

12. Hewitt and Seymour, "Factors," 102. American Association of University Women, *How Schools Shortchange Girls*, 56. Not only do women underestimate their chances of success, they often attribute success to things outside their control, such as luck. But when they perform poorly they tend to attribute their failure to a lack of ability. When men perform poorly, they tend to blame external factors, such as the difficult nature of the course materials or poor teaching.

13. Widnall, "AAAS Presidential Lecture," 1743. Dresselhaus: University of Minnesota, National Public Radio, "Science Lives," program no. 10 (1992). William Booth, "Oh, I Thought You Were a Man," *Science* 243 (27 Jan. 1989): 475. Georgina Ferry and Jane Moore, "True Confessions of Women in Science," *New Scientist* (July 1982). Sonnert and Holton, *Who Succeeds*, 145. Lifelong isolation and frustration can cast a long shadow. Between the years 1925 and 1979, one in ten women members of the American Chemical Society killed themselves: five times the national suicide rate for women of equivalent age. The rate of suicide among male chemists is only slightly higher than the national average. "Women Chemists Mortality Study Finds High Suicide Rate," *C&EN* 62 (23 April 1984).

14. Cole and Fiorentine, "Discrimination." Sonnert and Holton, *Gender Differences*, 144.

15. Norma Ware, Nicole Steckler, and Jane Leserman, "Undergraduate Women:

Who Chooses a Science Major?" *Journal of Higher Education* 56 (Jan./Feb. 1985); Zappert and Stanbury, "Pipeline"; Pearson, *Black Scientists;* N. Nevitte, R. Gibbins, and W.Codding, "The Career Goals of Female Science Students in Canada," *Canadian Journal of Higher Education* 18 (1988); Susan Frazier-Kouassi et al., *Women in Mathematics and Physics: Inhibitors and Enhancers* (University of Michigan: Center for the Education of Women, 1992). Sonnert and Holton, *Gender Differences,* 27.

16. A 1992 survey of the British Institute of Physics revealed that 58 percent of the women members had attended all-girls schools. Barinaga, "Surprises," 1472. Sebrechts, "Cultivating Scientists," 48. Susan Hill, *Undergraduate Origins of Recent Science and Engineering Doctorate Recipients* (Washington: NSF, 1992). Matyas and Malcom, eds., *Investing in Human Potential,* 15. Cheryl Leggon and Willie Pearson Jr., "The Baccalaureate Origins of African American Female Ph.D. Scientists," *Journal of Women and Minorities in Science and Engineering* 3 (1997).

17. Rosser, *Female-Friendly Science.* Falconer quoted in Sebrechts, "Cultivating Scientists," 48.

18. American Association of University Women, *Separated by Sex: A Critical Look at Single-Sex Education for Girls* (Washington: AAUW Educational Foundation, 1998).

19. Jean Kumagai, "Do Single-Sex Classes Help Girls Succeed in Physics," *Physics Today* 48 (Nov. 1995).

20. Committee on Women in Science and Engineering, *Women Scientists and Engineers,* 14; Anne Preston, "Why Have All the Women Gone?" *American Economic Review* 84 (1994).

21. Doreen Kimura, "Sex Differences in the Brain," *Scientific American* 267 (Sept. 1992): 125; for a review of this literature see Halpern, *Sex Differences.*

22. Martha Crouch, "Debating the Responsibilities of Plant Scientists in the Decade of the Environment," *Plant Cell* 2 (April 1990).

23. Regine Kollek, "Geschichte eines Ausstiegs," *Schritte ins Offene* 6 (1986). Similar discussions took place in the United States. Spanier, *Im/partial Science,* 125.

24. Matyas and Malcom, eds., *Investing in Human Potential,* 1–9.Committee on Women in Science and Engineering, *Women Scientists and Engineers,* 32.

4. The Clash of Cultures

1. M. G. Lord, *Forever Barbie: The Unauthorized Biography of a Real Doll* (New York: Morrow, 1994); Ann duCille, "Dyes and Dolls: Multicultural Barbie and the Merchandising of Difference," *Differences* 6 (1994); Jacqueline Urla and Alan Swedlund, "The Anthropometry of Barbie: Unsettling Ideals of the Feminine Body in Popular Culture," in *Deviant Bodies,* ed. Terry and Urla. There is now a Dr. Barbie, a pediatrician (still wearing ridiculously high heels), and an astronaut Barbie (last seen in boots). There is also a "caring careers" wardrobe featuring veterinarian's and firefighter's outfits. From the Nov. 1994 WISENET Barbie debate.

2. Karl Joël, *Die Frauen in der Philosophie* (Hamburg, 1896), 44, 48; Immanuel Kant, *Beobachtungen über das Gefühl des Schönen und Erhabenen*, in *Kants Werke*, ed. Wilhelm Dilthey (Berlin, 1900–1919), vol. 2, 229–230. Mary Wollstonecraft, *Vindication of the Rights of Woman* (1792), ed. Miriam Brody Kramnick (London: Penguin, 1982), 83.

3. Keller, *Reflections*. Barinaga, "Female Style."

4. Michel Foucault, *The Order of Things: An Archaeology of the Human Sciences* (1966; New York: Random House, 1970), xx; Snow, *Two Cultures*.

5. Martha Minow, "Learning to Live with the Dilemma of Difference: Bilingual and Special Education," *Law and Contemporary Problems* 48 (1984).

6. Marlene LeGates, "The Cult of Womanhood in Eighteenth-Century Thought," *Eighteenth-Century Studies* 10 (1976); Joan Landes, ed., *Feminism, the Public, and the Private* (Oxford: Oxford University Press, 1998).

7. G. W. F. Hegel, *Phänomenologie des Geistes* (1807), in *Werke*, ed. Eva Moldenhauer and Karl Michel (Frankfurt: Suhrkamp, 1969–1979), vol. 3, 319.

8. Rousseau, *Lettre à M. d'Alembert*, 152–155. Galton quoted in Easlea, "Masculine Image of Science," 137.

9. Georges Cuvier, "Extrait d'observations faites sur le cadavre d'une femme connue à Paris et à Londres sous le nom de Vénus Hottentotte," *Mémoires du Muséum d'Histoire Naturelle* 3 (1817): 272–273.

10. Margaret Mead and Rhoda Métraux, "Image of the Scientist among High-School Students," *Science* 126 (30 Aug. 1957).

11. Kahle, "Images of Science," 2–3. Deborah Fort and Heather Varney, "How Students See Scientists: Mostly Male, Mostly White, and Mostly Benevolent," *Science and Children* (May 1989): 12–13.

12. Irene Frieze and Barbara Hanusa, "Women Scientists: Overcoming Barriers," in *Advances in Motivation and Achievement*, ed. Steinkamp and Maehr, vol. 2, 145. LaFollette, *Making Science*, 66–77.Traweek, *Beamtimes*, 77–81.

13. Einstein, "Prinzipien der Forschung" (1918), quoted in Paul Forman, "Physics, Modernity, and our Flight from Responsibility," paper presented at the History of Science Society annual meeting, Santa Fe, Nov. 1993. Genevieve Lloyd, *The Man of Reason: "Male" and "Female" in Western Philosophy* (Minneapolis: University of Minnesota Press, 1984). Russell quoted in Easlea, "Masculine Image of Science," 136. David Brewster, *The Life of Sir Isaac Newton* (London, 1831), 341.

14. Nancy Tuana, "Revaluing Science: Starting from the Practices of Women," in *Feminism*, ed. Nelson and Nelson, 18.

15. Eve Curie, *Madame Curie: A Biography*, trans. Vincent Sheean (Garden City: Doubleday, 1937), 105–118."Interview with Andrea Dupree," 99.

16. Mary Beth Ruskai, "Why Women Are Discouraged from Becoming Scientists," *Scientist* (5 March 1990): 17. Elizabeth Pennisi, "Flexibility, Balance Draw Women to the University of Oregon," *Scientist* (15 Oct. 1990): 7. McGrayne, *Nobel Prize Women*, 48.

17. Anne Kinney, "Astronomizing at STScI," in *Women in Astronomy: Proceedings of a Workshop* (Baltimore: Space Telescope Science Institute, 1992), 194–195. The astronomer Laura Danly reported that men often rated her appear-

ance when she gave research talks. Diana Steel, "Astronomer Fights Sexism," *New Scientist* (26 Sept. 1992): 8. Linda Shepherd, *Lifting the Veil: The Feminine Face of Science* (Boston: Shambhala, 1993), 44. Henrion, *Women in Mathematics*, 73.

18. Watson, *Double Helix*, 14.

19. Hermann Weyl, "Emmy Noether," *Scripta Mathematica* 3 (July 1935): 219. Hyde et al., "Gender Comparisons," 310. Sime, *Lise Meitner*, 29, 33. Hubble quoted in Anne Eisenberg, "Women and the Discourse of Science," *Scientific American* (July 1992): 122.

20. Henley, *Body Politics*, 90. Susan Bordo, *Unbearable Weight: Feminism, Western Culture, and the Body* (Berkeley and Los Angeles: University of California Press, 1990). Kistiakowsky, "Women in Physics," 40.

21. Mark Whitaker, "White and Black Lies," *Newsweek* (15 Nov. 1993): 58.

22. Paul Kristeller, "Learned Women of Early Modern Italy: Humanists and University Scholars," in *Beyond Their Sex: Learned Women of the European Past*, ed. Patricia Labalme (New York: New York University Press), 102. Letter from Sophie Germain to C. F. Gauss, 20 Feb. 1807, *Oeuvres philosophiques de Sophie Germain*, ed. H. Stupuy (Paris, 1896), 271. Kenneth Manning, "The Complexion of Science," *Technology Review* (Nov./Dec. 1991): 63. Blackwell, *Opening the Medical Profession to Women*, vii.

23. Christopher Zeeman, "Private Games," in *A Passion for Science*, ed. Wolpert and Richards, 53.

24. Margenau et al., eds., *The Scientist*, 185. Richard Feynman, "The Development of the Space-Time View of Quantum Electrodynamics," *Science* 153 (12 Aug. 1966): 700, 708. John Randall Jr., *The Career of Philosophy* (New York: Columbia University Press, 1962), vol. 1, 4.

25. Rayman and Brett, *Pathways for Women*, 31. Girls are often portrayed as the naive science student. Robert Gilmore's 1994 *Alice in Quantumland* features a young, ballet slipper–shod Alice who is instructed in elementary particles by a bespectacled male figure (Wilmslow, England: Sigma Press, 1994). Paul Clark's 1938 *Alice in Virusland* served as his presidential address to the Society of American Bacteriologists. Clark's young, naive Alice was swept away into virusland while waiting for her father in his lab. Clark noted that his wife's name was also Alice and that she worked with him in the laboratory. I thank Maria Marco for drawing this source to my attention.

26. Betty Friedan, "Twenty Years after the Feminine Mystique," *New York Times Magazine* (27 Feb. 1983): 56.

27. Roy Miller, *The Japanese Language* (Chicago: University of Chicago Press, 1967), 289–290. Robin Lakoff, *Language and Woman's Place* (New York: Harper and Row, 1975).

28. Hilary du Cros and Laurajane Smith, "Why a Feminist Critique of Archaeology," in *Women in Archeology*, ed. Du Cros and Smith, xviii.

29. Tannen, *You Just Don't Understand*, 188–215. Sonnert and Holton, *Who Succeeds*, 143. Lawrence Rifkind and Loretta Harper, "Cross-gender Immediacy Behaviors and Sexual Harassment in the Workplace: A Communication Paradox," *IEEE Transactions on Professional Communication* 35 (Dec.

1992): 239; Judith Hall, *Nonverbal Sex Differences: Communication Accuracy and Expressive Style* (Baltimore: Johns Hopkins Press, 1984).

30. Tannen, *You Just Don't Understand*. Henley, *Body Politics*, 69.

31. Hochschild, *Second Shift*, 88. Arlie Hochschild, *The Managed Heart: Commercialization of Human Feeling* (Berkeley: University of California Press, 1983), 235.

32. Henley, *Body Politics*, 38.

33. William Bielby, "Sex Differences in Careers: Is Science a Special Case?" in *Outer Circle*, ed. Zuckerman et al., 1184–185.

34. "Interview with Andrea Dupree," 98.

35. In Sonnert and Holton's study of elite men and women scientists, nearly half of the men and more than half of the women interacted, at least sometimes, differently with male and female colleagues. Most women preferred interacting with women. Some women felt that their exclusion from informal contacts with men hindered their careers. *Who Succeeds*, 142–143.

36. Bernice Sandler, "Women Faculty at Work in the Classroom; or, Why It Still Hurts to Be a Woman in Labor," Center for Women Policy Studies, May 1993; Susan Basow, "Student Ratings of Professors Are Not Gender Blind," *Association for Women in Mathematics Newsletter* 24 (Sept./Oct. 1994): 20–21.

37. Ronald Hoy, "A 'Model Minority' Speaks Out on Cultural Shyness," *Science* 262 (12 Nov. 1993): 1117–18. Kathryn Knecht, letter to the editor, *Science* 261 (23 July 1993): 409 (slightly modified).

38. Evelynn Hammonds, "Race, Gender and the History of Women in Science," paper presented at the History of Science Society annual meeting, Santa Fe, Nov. 1993.

39. Zappert and Stanbury, "Pipeline," 18.

40. Yolanda Moses, *Black Women in Academe: Issues and Strategies* (Washington: Association of American Colleges, 1989); Adrianne Andrews, "Balancing the Personal and Professional," in *Spirit, Space and Survival: African American Women in (White) Academe*, ed. Joy James and Ruth Farmer (New York: Routledge, 1993), 189–190; Johnnetta Cole, keynote address, Black Women in the Academy Conference, MIT, Jan. 1994.

41. Robert Merton, *The Sociology of Science: Theoretical and Empirical Investigations* (University of Chicago Press, 1973). "Report of Working Group on Macho-ness," "Women in Physics," Aspen Center for Physics, July 1994; see also Catherine Kallin, Katherine Freese, and Elizabeth Simmons, "Aspen Focal Week on Women in Physics," *Gazette: A Newsletter of the Committee on the Status of Women in Physics of the American Physical Society* 15 (1995): 6–8.

42. Traweek, *Beamtimes*, 87–88. Pagels quoted in Easlea, "Masculine Image of Science," 135. Watson, *Double Helix*, 9, 25.

43. Ellis Sandoz, *The Vögelinian Revolution* (Baton Rouge: Louisiana State University Press, 1981), 37. I thank Malachi Hacohen for this information.

44. Traweek, *Beamtimes*, 89. Faye Flam, "Italy: Warm Climate for Women on the Mediterranean," *Science* 263 (11 March 1994): 1480–81.

45. Hewitt and Seymour, *Factors*, 60–64. Margaret Zerega and Herbert Walberg,

"School Science and Femininity," in *Advances in Motivation and Achievement*, ed. Steinkamp and Maehr, 43–44.

46. Interview with author, aired on WPSU Radio, 29 Oct. 1992.
47. Kenschaft and Keith, eds., *Winning Women into Mathematics*, 14. Interview with author, aired on WPSU Radio, 26 Nov. 1992.
48. Sharon Traweek, "Cultural Differences in High-Energy Physics: Contrasts between Japan and the United States," in *The "Racial" Economy of Science: Toward a Democratic Future*, ed. Sandra Harding (Bloomington: Indiana University Press, 1993), 401.
49. Novalis quoted in Edgar Zilsel, "Die Gesellschaftlichen Würzeln der romantischen Ideologie," *Der Kampf* 26 (1933): 154. In France it was during the period from the 1750s to the 1790s that scientists first tried to dissociate themselves from the literati. Wolf Lepenies, *Das Ende der Naturgeschichte: Wandel kultureller Selbstverständlichkeiten in den Wissenschaften des 18. und 19. Jahrhunderts* (Frankfurt: Suhrkamp, 1978).
50. Lambert, *Reflections nouvelles sur les femmes*, 132.
51. Ibid., 110–111; Lougee, *Le Paradis des Femmes*, 53. Roger Hahn, *The Anatomy of a Scientific Institution: The Paris Academy of Science, 1666–1803* (Berkeley: University of California Press, 1971).
52. Rousseau, *Lettre à M. d'Alembert*, 156–157. Martin Rudwick, *The Great Devonian Controversy: The Shaping of Scientific Knowledge among Gentlemanly Specialists* (Chicago: University of Chicago Press, 1985), 435–438.
53. Antoine Lavoisier, *Elements of Chemistry*, trans. Robert Kerr, in *Lavoisier, Fourier, Faraday*, ed. Robert Maynard Hutchins (Chicago: W. Benton, 1952), 1. Wilda Anderson, *Between the Library and the Laboratory: The Language of Chemistry in Eighteenth Century France* (Baltimore: Johns Hopkins University Press, 1984), 141.
54. Bruno Latour and Steven Woolgar, *Laboratory Life: The Construction of Scientific Facts* (Princeton: Princeton University Press, 1986), 229. Richard Lewontin, "'Honest Jim' Watson's Big Thriller about DNA," in *The Double Helix: Text, Commentary, Reviews, Original Papers*, ed. Gunther Stent (New York: Norton, 1980), 187.
55. Fedigan, "Science and the Successful Female."

5. Science and Private Life

1. Vetter, "Glass Ceiling," 13. Wallis, "Why a Curriculum in Women's Health," 57.
2. Lerner, *Creation of Feminist Consciousness*, 11.
3. I discuss heterosexual households because they are what has been studied. There is a good bit of literature on gay and lesbian families, lesbian couples, gay fathers, gays in the workplace, but very little looks at partnership and parenting issues in relation to science careers. This is an area ripe for study. See Louis Diamant, ed., *Homosexual Issues in the Workplace* (Washington: Taylor and Francis, 1993); Anthony R. D'Augelli and Charlotte J. Patterson, eds., *Lesbian, Gay, and Bisexual Identities over the Lifespan: Psychological*

Perspectives (New York: Oxford University Press, 1995); Ritch Savin-Williams and Kenneth Cohen, eds., *The Lives of Lesbians, Gays, and Bisexuals: Children to Adults* (Fort Worth: Harcourt Brace, 1996).

4. Ellen Galinsky, *National Study of the Changing Workforce* (New York: Families and Work Institute, 1993), 49, 51, 54. Hochschild, *Second Shift*, 3. *Centre Daily Times* (22 Jan. 1994).

5. Badinter, *Mother Love*.

6. Rayman and Brett, *Pathways for Women*, 31, 91.

7. David Noble, *A World without Women: The Christian Clerical Culture of Western Science* (New York: Knopf, 1992).

8. Rossiter, *Women Scientists* (1982), 15–16.

9. Cole and Zuckerman, "Marriage," 170.

10. Ann Gibbons, "Key Issue: Two-Career Science Marriage," *Science* 255 (13 March 1992): 1380. Mary Raffalli, "Why So Few Women Physicists?" *New York Times* (9 Jan. 1994): 28.

11. Long, "Measures of Sex Differences," 169. Pearson, *Black Scientists*, 147–148; Ivan Amato, "Profile of a Field—Chemistry: Women Have Extra Hoops to Jump Through," *Science* 255 (13 March 1992): 1373; Collins, *Black Feminist Thought*, 61.

12. Letters to the editor, *Science* 263 (11 March 1994): 1357. Aldhous, "Germany," 1476. In the United States a lab director at NIH suggested that one of his scientists abort her fetus because having a child might interfere with her research. Jocelyn Kaiser, "NIH Case Ends with Mysteries Unsolved," *Science* 277 (26 Sept. 1997): 1920.

13. Ann Gibbons, "Key Issue: Two-Career Science Marriage," *Science* 255 (13 March 1992): 1380.

14. Arlie Hochschild, *The Time Bind: When Work Becomes Home and Home Becomes Work* (New York: Holt, 1997).

15. Lerner, *Creation of Feminist Consciousness*, 11.

16. Cole and Zuckerman, "Marriage," 169. Norman Goodman, Edward Royce, Hanan Selvin, and Eugene Weinstein, "The Academic Couple in Sociology: Managing Greedy Institutions," in *Conflict and Consensus: A Festschrift in Honor of Lewis A. Coser*, ed. Walter Powell and Richard Robbins (New York: Free Press, 1984). Quotation from Linda Gardiner, *Emilie du Châtelet* (Wellesley College Center for Research on Women, typescript), ch. 1.

17. Hochschild, *Second Shift*. John Snarey, *How Fathers Care for the Next Generation: A Four-Decade Study* (Cambridge, Mass.: Harvard University Press, 1993), 34. James Bond, Ellen Galinsky, and Jennifer Swanberg, *The 1997 National Study of the Changing Workforce* (New York: Families and Work Institute, 1998), 38.

18. Ronni Sandroff, "When Women Make More Than Men," *Working Woman* (Jan. 1994): 41.

19. Larry May and Robert Strikwerda, "Fatherhood and Nurturance," in *Rethinking Masculinity*, ed. May and Strikwerda (Lanham, Md.: Rowman and Littlefield, 1992). Jerry Adler, "Building a Better Day," *Newsweek* (17 June

1996). "Working Moms and the Daddy Penalty," *U.S. News & World Report* (24 Oct. 1994). *New York Times* (29 Oct. 1995): 14.

20. Gerald Marwill, Rachel Rosenfeld, and Seymour Spilerman, "Geographic Constraints on Women's Careers in Academia," *Science* 205 (21 Sept. 1979); Spector, "Women Astronomers"; Richard Primack and Virginia O'Leary, "Cumulative Disadvantages in the Careers of Women Ecologists," *BioScience* 43 (March 1993).

21. Healy, "Women in Science." Robert Crease, "Canadian Chemist Takes on Working Women," *Science* 255 (28 Feb. 1992).

22. Other reasons included: men resent women colleagues (20 percent), parents discourage daughters (14 percent), inadequate skills (11 percent), wanting part time work (11 percent), science is a male-dominated field (7 percent), inappropriate career for women (7 percent), unfeminine (7 percent), restricts chances to marry (3 percent), and cannot succeed (0.7 percent). Alice Rossi, "Barriers to the Career Choice of Engineering, Medicine, or Science among American Women," in *Women and the Scientific Professions,* ed. Jacqueline Mattfeld and Carol Van Akens (Cambridge: MIT Press, 1965); Carolyn Stout Morgan, "College Students' Perceptions of Barriers to Women in Science and Engineering," *Youth and Society* 24 (Dec. 1992): 231.

23. Zappert and Stanbury, "Pipeline," 13. Sonnert and Holton, " 'Glass Ceiling,' " 8.

24. Ajzenberg-Selove, *A Matter of Choices,* 114–115. For most of her professional life Goeppert Mayer taught courses, did research, and supervised doctoral theses—all without pay. She was given a full-time professorship (at the University of California, San Diego) only after the publication of her Nobel Prize–winning work (she shared the Nobel Prize with Eugene Wigner in 1963). Rossiter, *Women Scientists* (1982), 195; Rossiter, *Women Scientists* (1995), 122–148.

25. In 1986, it was estimated that 700,000 couples in the general population commuted. Scott Heller, " 'Commuter Marriages' a Growing Necessity for Many Couples in Academe," *Chronicle of Higher Education* (22 Jan. 1986): 3.

26. Mercedes Foster, "A Spouse Employment Program," *BioScience* 43 (April 1993).

27. Jane Lubchenco and Bruce Menge, "Split Positions Can Provide a Sane Career Track: A Personal Account," *BioScience* 43 (April 1993).

28. Deborah Goldberg and Ann Sakai, "Career Options for Dual-Career Couples," *Bulletin of the Ecological Society of America* 74 (June 1993).

6. Medicine

1. Keller, *Secrets,* 78.

2. Primmer, "Women's Health Research," 302. Healy, "Women's Health," 566.

3. The term is Cynthia Russett's: Sexual Science: The Victorian Construction of Womanhood (Cambridge, Mass.: Harvard University Press, 1989).

4. J. B. Saunders and C. D. O'Malley, eds., *The Anatomical Drawings of Andreas Vesalius* (New York: Bonanza, 1982), 222–223. This pattern persists today: see Mendelsohn et al., "Sex and Gender Bias."

5. Fritz Weindler, *Geschichte der Gynäkologisch-anatomischen Abbildung* (Dresden: Von Zahn & Jänsch, 1908), 41; Mary Niven Alston, "The Attitude of the Church towards Dissection before 1500," *Bulletin of the History of Medicine* 16 (1944); Kate Campbell Hurd-Mead, *A History of Women in Medicine* (Haddam, Conn.: Haddam Press, 1938), 358–359.

6. For hundreds of years midwives dominated women's health care. In the seventeenth and increasingly in the eighteenth century, man-midwives began encroaching on this ancient privilege, and by the nineteenth century university-trained obstetricians had taken over the more scientific (and lucrative) parts of childbirth. Many midwives were run out of business by the attempt to make birthing more dependent on university training in anatomy from which women were excluded. But few challenged midwives' right to work in the countryside or to treat the poor. Donnison, *Midwives;* Marland, ed., *Art of Midwifery;* Adrian Wilson, *The Making of Man Midwifery* (Cambridge, Mass.: Harvard University Press, 1995); Nina Gelbart, The King's Midwife (Berkeley: University of California Press, 1998).

7. See John Riddle, *Contraception and Abortion from the Ancient World to the Renaissance* (Cambridge, Mass.: Harvard University Press, 1992); John Riddle, *Eve's Herbs: A History of Contraceptive and Abortion in the West* (Cambridge, Mass.: Harvard University Press, 1997).

8. See M. C. Horowitz, "Aristotle and Woman," *Journal of the History of Biology* 9 (1976); Ian Maclean, *The Renaissance Notion of Woman* (Cambridge: Cambridge University Press, 1980); Danielle Jacquart and Claude Thomasset, *Sexuality and Medicine in the Middle Ages,* trans. Matthew Adamson (Princeton: Princeton University Press, 1988); Joan Cadden, *Meanings of Sex Difference in the Middle Ages: Medicine, Science, and Culture* (Cambridge: Cambridge University Press, 1993).

9. Helkiah Crooke, *Mikrokosmographia: A Description of the Body of Man* (London, 1615), 249; Galen, *On the Usefulness of the Parts of the Body,* trans. Margaret May (Ithaca: Cornell University Press, 1968), vol. 2, 628–629.

10. François Poullain de la Barre, *De l'égalité des deux sexes: discours physique et moral* (Paris, 1673). Samuel Thomas von Soemmerring, *Über die körperliche Verschiedenheit des Negers vom Europäer* (Frankfurt and Mainz, 1785), preface.

11. Laqueur, *Making Sex;* Claudia Honegger, *Die Ordnung der Geschlechter: Die Wissenschaften vom Menschen und das Weib* (Frankfurt: Campus Verlag, 1992). Clarke, *Sex in Education,* 15.

12. The German Jakob Ackermann was one of the few physicians concerned about the implications of difference for health care. He appended to his lengthy study of sexual difference observations on how bodily differences between men and women might require different treatment of illnesses (such as fever). Jakob Ackermann, *Über die körperliche Verschiedenheit des Man-*

nes vom Weibe außer Geschlechtstheilen, trans. Joseph Wenzel (Koblenz, 1788). Clarke, *Sex in Education,* 33, 39, 62, 101–102, 136.

13. Sharon Begley, "Gray Matters," *Newsweek* (27 March 1995). Janice Irvine, "From Difference to Sameness: Gender Ideology in Sexual Science," *Journal of Sex Research* 27 (1990).

14. Teresa Ruiz and Lois Verbrugge, "A Two Way View of Gender Bias in Medicine," *Journal of Epidemiological Community Health* 51 (1997); Narrigan et al., "Research to Improve Women's Health"; Krieger and Fee, "Man-Made Medicine," 12–16.

15. Rosser, *Women's Health,* 6; Mastroianni et al., eds., *Women and Health Research;* Trisha Gura, "Estrogen: Key Player in Heart Disease among Women," *Science* 269 (11 Aug. 1995): 771.

16. A follow-up randomized trial was approved by NIH in 1991. Johnson and Fee, "Women's Health Research," 4–5.

17. Steering Committee of the Physicians' Health Study Research Group, "Final Report on the Aspirin Component of the On-Going Physicians' Health Study," *New England Journal of Medicine* 321 (1989); Office of Research on Women's Health, *Report of the National Institutes of Health,* 66; Linda Sherman, Robert Temple, and Ruth Merkatz, "Women in Clinical Trials: A FDA Perspective," *Science* 269 (11 Aug. 1995).

18. Barbara Rice, "Equity, Health Issues Should Define Women's Participation in Drug Studies," *AWIS Magazine* 23 (Sept./Oct. 1994): 14; Council on Ethical and Judicial Affairs, American Medical Association, "Gender Disparities in Clinical Decision Making," *JAMA* 266 (1991): 559.

19. Nechas and Foley, *Unequal Treatment,* 26.

20. Tracy Johnson and Elizabeth Fee, "Women's Participation in Clinical Research: From Protectionism to Access," in *Women and Health Research,* ed. Mastroianni et al., 5. Judy Norsigian, "Women and National Health Care Reform," *Journal of Women's Health* 2 (1993): 91.

21. Gamble and Blustein, "Racial Differentials in Medical Care," 184–187.

22. The FDA has set guidelines, but the inclusion of women in drug tests is not required by law. Ruth Merkatz and Elyse Summers, "Including Women in Clinical Trials: Policy Changes at the Food and Drug Administration," in *Women's Health Research,* ed. Haseltine and Jacobson.

23. Narrigan et al., "Research to Improve Women's Health," 564.

24. Florence Haseltine, "Foreword," in *Women's Health Research,* ed. Haseltine and Jacobson.

25. Haseltine and Jacobson, eds., *Women's Health Research.*

26. "Women Not Shortchanged in Trials?" *Science* 275 (14 March 1997). Office of Research on Women's Health, *Report of the National Health Institutes,* 8. Charles Mann, "Women's Health Research Blossoms," *Science* 269 (11 Aug. 1995). It is impossible to document conclusively whether women have been systematically excluded from clinical trials because NIH has not collected the necessary data. In articles between 1960 and 1991 in English-language journals, only 20 percent of research subjects in clinical trials of drugs for acute

heart attack were women. Women were also underrepresented in mixed-sex medical experiments unrelated to heart disease. The number of experiments on female-only and male-only disease has been about equal, though data were commonly underanalyzed for sexual differences. Women-only studies focused primarily on pregnancy and childbirth. Chloe Bird, "Women's Representation as Subjects in Clinical Studies," in *Women and Health Research*, ed. Mastroianni et al.

27. Nechas and Foley, *Unequal Treatment*, 227. Rachel Nowak, "New Push to Reduce Maternal Mortality in Poor Countries," *Science* 269 (11 Aug. 1995).

28. Krieger and Fee, "Man-Made Medicine"; Doyal, *What Makes Women Sick*; Ruzek et al., "Social, Biomedical, and Feminist Models."

29. A community approach to health may be fostered by a little-known aspect of the Women's Health Initiative: the Community Prevention Study, aimed at evaluating the community health practices of poor women. Many aspects of long-term health for women do not rely on clinical research but on access to medical care, healthy living, and information about birth control, the dangers of smoking, the benefits of exercise, and so forth. Communication from the Department of Health and Human Services, 19 May 1995.

30. Nancy Krieger and Stephen Sidney, "Racial Discrimination and Blood Pressure: The CARDIA Study of Young Black and White Adults," *American Journal of Public Health* 86 (1996): 1375; Diane Adams, ed., *Health Issues for Women of Color: A Cultural Diversity Perspective* (Thousand Oaks: Sage, 1995).

31. Ruzek et al., eds., *Women's Health*, xv. Deborah Wingard, "Patterns and Puzzles: The Distribution of Health and Illness among Women in the United States," in *Women's Health*, ed. Ruzek et al.

32. Krieger and Fee, "Man-Made Medicine," 18–19.

33. Ruzek et al., "Social, Biomedical, and Feminist Models," 15. See Nancy Krieger, Jarvis Chen, and Gregory Ebel, "Can We Monitor Socioeconomic Inequalities in Health? A Survey of U.S. Health Departments' Data Collection and Reporting Practices," *Public Health Reports* 112 (1997).

34. Karyn Montgomery and Anne Moulton, "Medical Education in Women's Health," *Journal of Women's Health* 1 (1992). I thank Sue Rosser for the information concerning the American College of Women's Health.

35. Susan Sperling and Yewoubdar Beyene, "A Pound of Biology and a Pinch of Culture or a Pinch of Biology and a Pound of Culture? The Necessity of Integrating Biology and Culture in Reproductive Studies," in *Women in Human Evolution*, ed. Hager.

36. Margaret Lock, *Encounters with Aging: Mythologies of Menopause in Japan and North America* (Berkeley: University of California Press, 1993).

37. Elizabeth Fee and Nancy Krieger, "Introduction," in *Women's Health, Politics, and Power*, ed. Fee and Krieger, 1–2.

38. Evelyn White, *Black Women's Health Book* (Seattle: Seal Press, 1990); Rosser, *Women's Health*; Vivian Pinn, *Overview: Office of Research on Women's Health* (Bethesda: NIH, 1995); Shirley Malcom, "Science and Diversity: A Compelling National Interest," *Science* 271 (29 March 1996).

39. See, e.g., Birke, *Women, Feminism, and Biology;* Martin, *Woman in the Body;* Hubbard, *The Politics of Women's Biology.*
40. Mendelsohn et al., "Sex and Gender Bias."
41. Primmer, "Women's Health Research."
42. National Institutes of Health, *Problems in Implementing Policy on Women in Study Populations* (Washington: General Accounting Office, National and Public Health Issues and Human Resources Division, June 1990). Primmer, "Women's Health Research," 303.
43. Johnson and Fee, "Women's Health Research," 17. Florence Haseltine, " Formula for Change: Examining the Glass Ceiling," in *Women's Health Research,* ed. Haseltine and Jacobson; Lillian Randolph, Bradley Seidman, and Thomas Pasko, *Physician Characteristics and Distribution in the United States* (Chicago: American Medical Association, 1997). Women are clustered in the lower ranks of medical school faculties: 50 percent of assistant professors, 20 percent of associate professors, and 10 percent of full professors are women. Tracy Johnson and Susan Blumenthal, "Women in Academic Medicine," *Journal of Women's Health* 2 (1993): 216. Moving women up in the academic hierarchy is difficult. Perhaps a national effort such as the one being undertaken by the Office of Research on Women's Health can succeed where others have failed. Toward this end, the ORWH produced a report, "Women in Biomedical Careers: Dynamics of Change, Strategies for the Twenty-First Century" (1994), outlining issues in education, career advancement, and social barriers along with strategies for improvement.

7. Primatology, Archaeology, and Human Origins

1. Johanson and Edey, *Lucy,* 18, 269. Lori Hager, "Sex and Gender in Paleoanthropology," in *Women in Human Evolution,* ed. Hager.
2. Ian Tattersall, *The Human Odyssey: Four Million Years of Human Evolution* (New York: Prentice Hall, 1993), 75–76.
3. I thank Trudy Turner and Linda Fedigan for these numbers. Jeffrey French, "A Demographic Analysis of the Membership of the American Society of Primatologists: 1992," *American Journal of Primatology* 29 (1993); Fedigan, "Science and the Successful Female."
4. Fedigan, "Changing Role of Women."
5. Fedigan and Fedigan, "Gender and the Study of Primates," 41.
6. Fedigan, "Changing Role of Women," 39.
7. Strum and Fedigan, "Theory, Method and Gender."
8. Jane Lancaster, "In Praise of the Achieving Female Monkey," *Psychology Today* 7 (1973). Early classics on female primates include Jeanne Altmann, *Baboon Mothers and Infants* (Cambridge, Mass.: Harvard University Press, 1980); Hrdy, *The Woman That Never Evolved;* Linda Fedigan, *Primate Paradigms: Sex Roles and Social Bonds* (Montreal: Eden Press, 1982); Meredith Small, ed., *Female Primates: Studies by Women Primatologists* (New York: Alan Liss, 1984); Barbara Smuts, *Sex and Friendship in Baboons* (New York: Aldine, 1985); Jane Goodall, *Chimpanzees of Gombe: Patterns of Behavior*

(Cambridge, Mass.: Harvard University Press, 1986); Shirley Strum, *Almost Human* (New York: Random House, 1987).

9. Rowell, "Introduction," 16. Strum and Fedigan, "Theory, Method and Gender."

10. Edward O. Wilson, *Sociobiology: The New Synthesis* (Cambridge, Mass.: Harvard University Press, 1975), 553. Hrdy, *The Woman That Never Evolved;* Haraway also numbers Barbara Smuts among the first feminist sociobiologists; see *Primate Visions.*

11. Sarah Hrdy and G. Williams, "Behavioral Biology and the Double Standard," in *Social Behavior of Female Vertebrates,* ed. Samuel Wasser (New York: Academic Press, 1983), 7. Darwin, *Descent,* pt. 2, 312–315. Marcy Lawton, William Garstka, and Craig Hanks, "The Mask of Theory and the Face of Nature," in *Feminism and Evolutionary Biology,* ed. Gowaty.

12. Jeanne Altmann, "Mate Choice and Intersexual Reproductive Competition: Contributions to Reproduction That Go Beyond Acquiring More Mates," in *Feminism and Evolutionary Biology,* ed. Gowaty, 329; Patricia Gowaty, "Sexual Dialectics, Sexual Selection, and Variation in Reproductive Behavior," ibid. Hrdy, "Empathy"; Judy Stamps, "Role of Females," 302–308.

13. Susan Sperling, "Baboons with Briefcases vs. Langurs in Lipstick: Feminism and Functionalism in Primate Studies," in *Gender at the Crossroads,* ed. di Leonardo, 218.

14. Longino citing Haraway and Sperling, "Cognitive and Non-Cognitive Values," 52. Zihlman, "Misreading Darwin," 431–432. Gowaty, "Introduction," in *Feminism and Evolutionary Biology,* ed. Gowaty, 7. Fedigan, "Is Primatology a Feminist Science?"

15. See, e.g., Linda Birke, *Feminism, Animals, and Science: The Naming of the Shrew* (Buckingham: Open University Press, 1994); *Hypatia* 6 (1991) (a special issue on ecological feminism).

16. In Carolyn Merchant's taxonomy in *Earthcare,* my critique applies only to cultural ecofeminism.

17. Strum and Fedigan, "Theory, Method and Gender."

18. Gross and Levitt, *Higher Superstition.*

19. Richard Wrangham, "Subtle, Secret Female Chimpanzees," *Science* 277 (8 Aug. 1997).

20. See, e.g., Lee and DeVore, eds., *Man the Hunter;* Fedigan, "Changing Role of Women," 29, 32–33; Balme and Beck, "Archaeology and Feminism," 63.

21. Zihlman, "Paleolithic Glass Ceiling"; Adrienne Zihlman, "Woman the Gatherer: The Role of Women in Early Hominid Evolution," in *Gender and Anthropology,* ed. Morgen; Fedigan, "Changing Role of Women."

22. Zihlman, "Sex, Sexes, and Sexism," 14; Zihlman, "Paleolithic Glass Ceiling," 95–96, 98.

23. Haraway, *Primate Visions,* 334. Zihlman, "Sex, Sexes, and Sexism," 13.

24. Balme and Beck, "Archaeology and Feminism." Conkey with Williams, "Original Narratives," 114, 123.

25. Zihlman, "Paleolithic Glass Ceiling," 100–103; Zihlman, "Misreading Darwin," 436.

26. Quoted in Nelson et al., eds., *Equity Issues for Women in Archeology,* 5.
27. Conkey, "Making the Connections," 3. Bacus et al., eds., *Gendered Past.* Gero and Conkey, eds., *Engendering Archaeology;* Morgen, ed., *Gender and Anthropology;* Nelson et al., eds., *Equity Issues for Women in Archeology;* Cheryl Claassen, ed., *Women in Archaeology* (Philadelphia: University of Pennsylvania Press, 1994); Alison Wylie, Margaret Conkey, and Ruth Trignham, "Archaeology and the Goddess: Exploring the Contours of Feminist Archaeology," in *Feminisms in the Academy,* ed. Domna Stanton and Abigail Stewart (Ann Arbor: University of Michigan Press, 1995); Cheryl Claassen and Rosemary Joyce, eds., *Women in Prehistory: North America and Mesoamerica* (Philadelphia: University of Pennsylvania Press, 1997).
28. Conkey with Williams, "Original Narratives."
29. William Laughlin, "Hunting: An Integrating Biobehavior System and Its Evolutionary Importance," in *Man the Hunter,* ed. Lee and DeVore, 318. Conkey, "Making the Connections," 11.
30. Joan Gero, "Excavation Bias and the Woman at Home Ideology," in *Equity Issues for Women in Archeology,* ed. Nelson et al.
31. Joan Gero, "Genderlithics: Women's Roles in Stone Tool Production," in *Engendering Archaeology,* ed. Gero and Conkey. Gero, "Social World of Prehistoric Facts." Half of those engaged in micro-wear studies (studying flakes for evidence of use) are women, though women make up only 20 percent of archaeologists in North America.
32. Rita Wright, "Women's Labor and Pottery Production in Prehistory," in *Engendering Archaeology,* ed. Gero and Conkey.
33. Conkey and Spector, "Archaeology and the Study of Gender," 11.
34. Patty Jo Watson and Mary Kennedy, "The Development of Horticulture in the Eastern Woodlands of North America: Women's Role," in *Engendering Archaeology,* ed. Gero and Conkey, 262.
35. Wylie, "Engendering of Archaeology."
36. Ibid., 96.

8. Biology

1. Marian Lowe, "The Impact of Feminism on the Natural Sciences," in *The Knowledge Explosion: Generations of Feminist Scholarship,* ed. Cheris Kramarae and Dale Spender (New York: Teachers College Press, 1992), 162.
2. Biology and Gender Study Group, "Importance of Feminist Critique"; Martin, "Egg and Sperm"; Keller, *Refiguring Life,* xii–xiii.
3. Gerald Schatten and Heide Schatten, "The Energetic Egg," *Sciences* 23 (Sept/Oct 1983). Spanier, *Im/partial Science,* 60.
4. Biology and Gender Study Group, "Importance of Feminist Critique," 172.
5. Martin, "Egg and Sperm," 498–499. Spanier, *Im/partial Science,* 62. Spanier notes that sociobiologists have taken the larger size of the egg to support the notion that females have a greater "parental investment" in their offspring, leading them to suggest that females ought to be the more engaged parents in caring for offspring.

6. Martin, "Egg and Sperm," 501.
7. Gross and Levitt, *Higher Superstition*, 121. Squier, *Babies in Bottles*.
8. Biology and Gender Study Group, "Importance of Feminist Critique"; Spanier, *Im/partial Science*, 62–63. Joseph Palca, "The Other Human Genome," *Science* 249 (7 Sept. 1990): 1104.
9. See Scott Gilbert, "Cellular Politics: Ernest Everett Just, Richard B. Goldschmidt, and the Attempt to Reconcile Embryology and Genetics," in *The American Development of Biology*, ed. Ronald Rainger, Keith Benson, and Jane Maienschein (Philadelphia: University of Pennsylvania Press, 1988); Keller, *Refiguring Life*, 36–40.
10. Spanier, *Im/partial Science*, 87–88.
11. William Smellie, *The Philosophy of Natural History*, 2 vols. (Edinburgh, 1790), vol. 1, 237, 238. Desfontaines quoted in François Delaporte, *Nature's Second Kingdom: Explorations of Vegetality in the Eighteenth Century*, trans. Arthur Goldhammer (Cambridge, Mass.: MIT Press, 1982), 129.
12. Nancy Marie Brown, "The Wild Mares of Assateague," *Research/Penn State* 16 (Dec. 1995); A. Innis Dagg, *Harems and Other Horrors in Behavioral Biology* (Waterloo, Ontario: Otter Press, 1983).
13. Roberta Bivins, "Sex and the Single Cell: Gender and the Language of Molecular Biology" (Program in Science, Technology, and Society, MIT, manuscript, 1997).
14. Lynn Margulis and Dorion Sagan, *Origins of Sex: Three Billion Years of Genetic Recombination* (New Haven: Yale University Press, 1986), 54–55. Spanier, *Im/partial Science*, 56–59.
15. James Darnell, Harvey Lodish, and David Baltimore, *Molecular Cell Biology* (New York: Scientific American Books, 1986), "corrected" in the 1990 ed. Keller, *Refiguring Life*, 24.
16. Quoted in Fausto-Sterling, "Life in the XY Corral," 327.
17. Anne Fausto-Sterling, "Society Writes Biology/Biology Constructs Gender," *Daedalus* 116 (1987); Hubbard, *Profitable Promises*, 169–170; Spanier, *Im/partial Science*, 70–72.
18. R. V. Short, "Sex Determination and Differentiation," in *Reproduction in Mammals*, ed. C. R. Austin and R. V. Short (Cambridge: Cambridge University Press, 1972), vol. 2, 70.
19. Eva Eicher and Linda Washburn, "Genetic Control of Primary Sex Determination in Mice," *Annual Review of Genetics* 20 (1986): 328–329. I thank Scott Gilbert for his comments on this section. Page quoted in Maya Pines, "Becoming a Male, Becoming a Female," in *From Egg to Adult* (Bethesda: Howard Hughes Medical Institute, 1992), 42–45.
20. This statement is intended rhetorically; social factors contributing to a shorter male life span are discussed in Chapter 6.
21. Bettyann Kevles, *Females of the Species: Sex and Survival in the Animal Kingdom* (Cambridge, Mass.: Harvard University Press, 1986), 201–203.
22. Longino, *Science as Social Knowledge*.
23. Ibid.
24. Badinter, *Mother Love*.

25. Keller, *Secrets*, 148.
26. Messer-Davidow et al., eds., *Knowledges*, preface.
27. On reductionism see, e.g., Hubbard, Henifin, and Fried, eds., *Biological Woman*; Bleier, *Science and Gender*; Birke and Hubbard, eds., *Reinventing Biology*. Watson, *Double Helix*, 19. Fausto-Sterling, "Life in the XY Corral"; Keller, Secrets.
28. Hubbard, *Profitable Promises*; Rose, *Love, Power, and Knowledge*, 204.

9. Physics and Math

1. Jean Kumagai, "Women See Gains in U.S. Physics Professoriat," *Physics Today* (Sept. 1994): 86. I thank Judith Mulven, Statistics Division, AIP, for her aid.
2. Isaac Newton, *Principes mathématiques de la philosophie naturelle*, trans. Marquise du Chastellet (Paris, 1756).
3. Harding, *Whose Science*. Barad, "A Feminist Approach to Teaching Quantum Physics." On value neutrality see Proctor, *Value-Free Science*; Bleier, ed., *Feminist Approaches to Science*; Keller, *Reflections*; Harding, *Science Question*; Schiebinger, *Mind*; Haraway, *Primate Visions*; Keller and Longino, eds., *Feminism and Science*.
4. Conkey, "Making the Connections," 3.
5. Julie Klein, "Blurring, Cracking, and Crossing: Permeation and the Fracturing of Disciplines," in *Knowledges*, ed. Messer-Davidow et al., 188. Zuckerman, "Careers," 31.
6. Bertrand Russell, *Our Knowledge of the External World* (New York: Norton, 1929), 75–79. I thank Robert Merton for calling this passage to my attention.
7. Holton quoted in Phil Allport, "Still Searching for the Holy Grail," *New Scientist* 132 (5 Oct. 1991): 56. Scott Gilbert, "Resurrecting the Body: Has Postmodernism Had Any Effect on Biology?" *Science in Context* 8 (1995): 568. Traweek, *Beamtimes*, 78–79. Stephen Brush, "Should the History of Science Be Rated X?" *Science* 183 (22 March 1974): 1164.
8. Morrow and Morrow, "Whose Math Is It," 50. Robert Westman, "Two Cultures or One? A Second Look at Kuhn's *The Copernican Revolution*," *Isis* 85 (1994): 92.
9. Martin Rees, "Contemplating the Cosmos," in *A Passion for Science*, ed. Wolpert and Richards, 34–35. Virginia Morell, "Rise and Fall of the Y Chromosome," *Science* 263 (14 Jan. 1994): 171.
10. Karen Barad, "Meeting the Universe Halfway: Realism and Social Constructivism without Contradiction," in *Feminism*, ed. Nelson and Nelson, 168–173.
11. Barad, "A Feminist Approach to Teaching Quantum Physics," 64.
12. Wilson quoted in Margaret Wertheim, *Pythagoras' Trousers: God, Physics, and the Gender Wars* (New York: Times Books, 1995), 220–221.
13. Traweek, *Beamtimes*, 104.
14. Galison and Hevly, eds., *Big Science*, preface.
15. Traweek, "Big Science." Galison, "The Many Faces of Big Science," 3. La-

Follette, *Making Science Our Own*, 11–12. Forman, "Behind Quantum Electronics," 152. Forman has documented the historic growth and qualitative change of American physics during the 1940s when it spearheaded the effort to provide national security through ever more advanced military technologies.

16. Forman, "Behind Quantum Electronics," 152–153. In 1985 the Department of Defense contributed 50 percent of the federal funding for university research in mathematics and computer science. William Hartung and Rosy Nimroody, "Star Wars: Pentagon Invades Academia," *Association for Women in Mathematics Newsletter* 17 (1987).

17. Jean Kumagai, "AIP Survey Finds More Women Majoring in Physics," *Physics Today* (July 1990): 64. Barton Reppert, "1995 Budget Draws Praise—and Concerns," *Scientist* 8 (31 Oct. 1994). Constance Holden, "Science Careers: Playing to Win," *Science* 265 (23 Sept. 1994). In 1994 unemployment among new math Ph.D.'s was 9 percent, and this statistic does not represent the numbers of young scientists in temporary posts or repeated postdocs.

18. Forman, "Behind Quantum Electronics," 156–157. Wim Smit, "Science, Technology, and the Military: Relations in Transition," in *Handbook of Science and Technology Studies*, ed. Jasanoff et al., 607–611.

19. Brian Easlea, *Fathering the Unthinkable: Masculinity, Scientists, and the Nuclear Arms Race* (London: Pluto, 1983); Carol Cohn, "Slick'ems, Glick'ems, Christmas Trees, and Cookie Cutters: Nuclear Language and How We Learned to Pat the Bomb," *Bulletin of the Atomic Scientists* 43 (June 1987), 20; Keller, *Secrets*, 44.

20. Daniel Kevles, *The Physicists* (New York: Knopf, 1978), 202. MIT president quoted in Galison, "Many Faces of Big Science," 8. Rossiter, *Women Scientists* (1995), 133. Women were sometimes encouraged to go into science in the 1920s and 1930s as part of the project to establish national strength in science. Ernest Rutherford wished to make Cambridge an "Imperial University." The loss of young men during the First World war required that he recruit women. "Women," he declared, "are often endowed with such a degree of intelligence as enables them to contribute substantially to progress in the various branches of learning; at the present stage in the world's affairs, we can afford less than ever to neglect the training and cultivation of all the young intelligence available." Teri Hopper, " 'Radioactive Ladies and Gentlemen': Women and Men of the Radioactivity Community, 1919–1939," paper presented at the History of Science Society Annual Meeting, 28 Oct. 1995.

21. On women entering the U.S. armed forces, see Linda Bird Francke, *Grand Zero: The Gender Wars in the Military* (New York: Simon and Schuster, 1997). Meitner quoted in Louis Haver, *Women Pioneers in Science* (New York: Harcourt Brace Jovanovich, 1979), 50. Also private communication from Ruth Sime (28 March 1994); see Sime, "Lise Meitner in Sweden 1938–1960: Exile from Physics," *American Journal of Physics* 62 (1994): 698, and *Lise Meitner*.

22. Jane Wilson and Charlotte Serber, eds., *Standing By and Making Do: Women*

of *Wartime Los Alamos* (Los Alamos: Los Alamos Historical Society, 1988). Galison, "Fortran," 229. Herzenberg and Howes, "Women of the Manhattan Project."

23. Herzenberg and Howes, "Women of the Manhattan Project," 38. Elaine May, *Homeward Bound: American Families in the Cold War Era* (New York: Basic Books, 1988), 103–104.

24. Debra Rosenthal, *At the Heart of the Bomb: The Dangerous Allure of Weapons Work* (Reading, Mass.: Addison-Wesley, 1990), 204–205. Gusterson, "Becoming a Weapons Scientist."

25. Gusterson, "Becoming a Weapons Scientist," 262.

26. Traweek, "Big Science," 102.

27. "Interview with Andrea Dupree," 103–105.

28. See Barbara Whitten, "What Physics Is Fundamental Physics? Feminist Implications of Physicists' Debate over the Superconducting Supercollider," *National Women's Studies Association Journal* 8 (1996).

29. Kumagai, "Survey and Site Visits," 57–59. Kistiakowsky, "Women in Physics," 38. Office of Research on Women's Health, *Summary: Public Hearing on Recruitment, Retention, Re-Entry, and Advancement of Women in Biomedical Careers* (Bethesda: NIH, 1992), 11.

30. "Women in Mathematics," *Science* 000 (17 July 1992): 323; Eleanor Babco and Betty Vetter, "Diversity of Women Scientists across Science Employment Sectors," *AWIS Magazine* 24 (Jan./Feb. 1995): 15. Henrion, *Women in Mathematics*.

31. Irving Kaplansky and John Riordan, "The Problème des Ménages," *Scripta Mathematica* 12 (1946). Kenneth Bogart and Peter Doyle, "Non-Sexist Solution of the Ménage Problem," *Mathematical Monthly* 93 (Aug./Sept. 1986). Kaplansky and Riordan report that of the many mathematicians to work on this problem, only one chose to seat the men first.

32. Keller, *Reflections on Gender and Science*.

33. Irigaray critiqued in N. Katherine Hayles, "Gender Encoding in Fluid Mechanics: Masculine Channels and Feminine Flows," *Differences* 4 (1992): 16–17.

34. Morrow and Morrow, "Whose Math Is It," 50. Hilary Lips, "Bifurcation of a Common Path: Gender Splitting on the Road to Engineering and Physical Science Careers," *Initiatives* 55 (1993).

35. J. Möbius, *Ueber die Anlage zur Mathematik* (Leipzig, 1900), 84–86. Anna Carlotte Leffler, *Sonya Kovalevsky: Her Recollections of Childhood*, trans. Isabel Hapgood and Clive Bayley (New York, 1895), 219.

36. Meredith Kimball, "A New Perspective on Women's Math Achievement," *Psychological Bulletin* 105 (1989): 199. NSF, *Women, Minorities* (1994), xxxii, 27–28. American Association of University Women, *How Schools Shortchange Girls*, 24–25.

37. Drawn from Anne Fausto-Sterling's excellent *Myths of Gender*, 13–60.

38. Suzanne Kavrell and Anne Petersen, "Patterns of Achievement in Early Adolescence," in *Women in Science*, ed. Steinkamp and Maehr. Halpern, *Sexual Differences*, 148–151, 163. Doreen Kimura argues that language differences

in men and women result from differences in posterior and anterior brain organization rather than in organization across or within hemispheres; Kimura, "Sex Differences in the Brain," *Scientific American* 267 (Sept. 1992).

39. See Sharon Begley, "Gray Matters," *Newsweek* (27 March 1995).

40. NSF, *Women, Minorities* (1994), 28. Ann Gallagher and Richard De Lisi, "Gender Differences in Scholastic Aptitude Test—Mathematics Problem Solving among High-Ability Students," *Journal of Educational Psychology* 86 (1994). Halpern, *Sex Differences*, 149.

41. Fausto-Sterling, *Myths of Gender*, 26–30. Quinn McNemar, *The Revision of the Stanford-Binet Scale: An Analysis of the Standardization Data* (Boston: Houghton Mifflin, 1942), 45. Rosser, *SAT Gender Gap;* Halpern, *Sex Differences*, 92–94.

42. Rosser, *SAT Gender Gap*, 52. Thomas Donlon, ed., *The College Board Technical Handbook for the Scholastic Aptitude Test and Achievement Tests* (New York: College Entrance Examination Board, 1984), 51–52.

43. Rosser, *SAT Gender Gap*, 55–56. In 1975 the math gender gap, which had hovered around 42 points, climbed to 50 points after the time allowed to work the test was shortened by 15 minutes and the data-sufficiency section (on which women scored higher) was eliminated. Rosser, *SAT Gender Gap: ETS Responds*, 5.

44. Hyde et al., "Gender Differences in Mathematics Performance"; Fennema and Leder, eds., *Mathematics and Gender*. Thomas Donlon, "Content Factors in Sex Differences on Test Questions," Research Memorandum 73–28 (Princeton: Educational Testing Service, 1973), cited in Phyllis Rosser, *SAT Gender Gap: ETS Responds*, 5. Betsy Becker, "Item Characteristics and Gender Differences on the SAT-M for Mathematically Able Youths," *American Educational Research Journal* 27 (Spring 1990); Rosser, *SAT Gender Gap*, 47–67.

45. Howard Wainer and Linda Steinberg, "Sex Differences in Performance on the Mathematics Section of the Scholastic Aptitude Test: A Bidirectional Validity Study," *Harvard Educational Review* 62 (1992). Hyde et al., "Gender Differences in Mathematics Performance"; Janet Hyde and Marcia Linn, "Gender Differences in Verbal Ability: A Meta-Analysis," *Psychological Bulletin* 105 (1988); Hyde et al., "Gender Comparisons of Mathematics Attitudes and Affects." Rosser, *SAT Gender Gap*, 4, 61, 87, 173–190, 56. Michael Behnke, testimony before the Congressional Subcommittee on Civil and Constitutional Rights, 23 April 1987.

46. Rosser, *SAT Gender Gap*, 61. American Association of University Women, *How Schools Shortchange Girls*, 52.

47. Alan Bayer and John Folger, "Some Correlates of a Citation Measure of Productivity in Science," *Sociology of Education* 39 (1966).

48. Camilla Benbow, "Sex Differences in Mathematical Reasoning Ability in Intellectually Talented Preadolescents," *Behavioral and Brain Sciences* 11 (1988): 182. American Association of University Women, *How Schools Shortchange Girls*, 26.

49. Brandon et al., "Children's Mathematics Achievement in Hawaii." M. M. Schratz, "A Developmental Investigation of Sex Differences in Spatial (Visual-

Analytical) and Mathematical Skills in Three Ethnic Groups," *Developmental Psychology* 14 (1978). Rosser, *SAT Gender Gap*, 50. Other studies, however, have shown that among American ethnic groups boys consistently out score girls, with Native Americans having the largest math gender gap and African Americans the smallest. Rosser, *SAT Gender Gap*, 57.

50. NSF, *Women, Minorities* (1994), 31. Rosser, *SAT Gender Gap*, 66.
51. Lilli Hornig, "Women Graduate Students: A Literature Review and Synthesis," in *Women*, ed. Dix, 111.
52. I thank Amy Bug for these examples and for her thoughtful contributions to my conclusions.

Conclusion

1. That physicists like Karen Barad and Amy Bug are working on this topic within physics departments is a new and encouraging development.
2. Peggy McIntosh, "Interactive Phases of Curricular Re-Vision: A Feminist Perspective," Working Paper no. 124, Wellesley College, Center for Research on Women, Oct. 1983. Sue Rosser and Linda Fedigan have modified this for an analysis of primatology: Rosser, "The Relationship between Women's Studies and Women in Science," in *Feminist Approaches to Science*, ed. Bleier; Fedigan, "Is Primatology a Feminist Science?" Merchant, *Earthcare*, 8; Schiebinger, "Creating Sustainable Science." Rose, "Hand, Brain, and Heart"; Rose, *Love, Power, and Knowledge*.
3. Haraway, *Simians*; Harding, *Whose Science*. See Robert Proctor's review of Donna Haraway's *Modest_Witness@Second_Millennium. FemaleMan©_ Meets_OncoMouse™* (New York: Routledge, 1997), *Bulletin of the History of Medicine* 72 (Summer 1998).
4. Tara McLoughlin, "CSWP Sponsors Site Visits Sessions," *Gazette: A Newsletter of the Committee on the Status of Women in Physics of the American Physical Society* 15 (Summer 1995); Kumagai, "Survey and Site Visits."
5. Introducing even minimal biographical materials on women scientists has been shown to have a positive effect on students' attitudes toward women's participation in science. Jill Marshall and James Dorward, "The Effect of Introducing Biographical Material on Women Scientists into the Introductory Physics Curriculum," *Journal of Women and Minorities in Science and Engineering* 3 (1997).
6. Rosser, ed., *Teaching the Majority* provides state-of-the-art methods for teaching traditional material in a way that is responsive to women and gender concerns. Biology and Gender Study Group, "Importance of Feminist Critique." Gilbert, *Developmental Biology*. Martin, *Woman in the Body*, xii.
7. Biology and Gender Study Group, "Importance of Feminist Critique," 172–173.
8. Fedigan, "Changing Role of Women."
9. Alison Wylie calls this the "integrative critique": seeing how the place of women in a discipline molds knowledge in that discipline. Wylie, "Engendering of Archaeology."

10. Longino, "Subjects."
11. Squier, *Babies in Bottles*. Keller, *Secrets*, 27–28.
12. Conkey, "Making the Connections"; Gero, "Social World of Prehistoric Facts."
13. Elisabeth Lloyd, "Pre-Theoretical Assumptions in Evolutionary Explanations of Female Sexuality," in *Feminism and Science*, ed. Keller and Longino, 96.
14. Martha McCaughey, "Perverting Evolutionary Narratives of Heterosexual Masculinities," *GLQ: A Journal of Lesbian and Gay Studies* 3 (1996).
15. Cole and Fiorentine, "Discrimination against Women in Science," 205. Messer-Davidow et al., eds., *Knowledges*. Lynn Doering, "Power and Knowledge in Nursing: A Feminist Poststructuralist View," *Advances in Nursing Science* 14 (1992): 27–28.
16. *Science* 269 (11 Aug. 1995): 773. Private communication from Catherine Kallin, Department of Physics, McMaster University. Mary Clutter, "Support of Conferences, Meetings, Workshops, and International Congresses," NSF/AD/BBS Circular no. 14 (15 Oct. 1991); Brigid Hogan, "Women in Science," *Nature* 360 (19 Nov. 1992): 204.
17. See Rosser, *Re-Engineering Female Friendly Science*.
18. Pressure from various government departments forced the watering down of this report. Britain, if moving ahead, is doing so slowly.
19. Niedersächsisches Ministerium für Wissenschaft und Kultur, *Berichte*. Despite these initiatives, much remains to be done in Germany to integrate women's studies into universities. Nigel Williams, "EU Moves to Decrease the Gender Gap," *Science* 280 (8 May 1998): 822.
20. Rachel Weber, "Manufacturing Gender in Commercial and Military Cockpit Design," *Science, Technology, and Human Values* 22 (1997).

Bibliography

Works included in the bibliography receive a short citation in the notes; works not included in the bibliography are fully cited in the notes.

Abir-Am, Pnina, and Dorinda Outram, eds. *Uneasy Careers and Intimate Lives: Women in Science, 1789–1979.* New Brunswick: Rutgers University Press, 1987.

Agarwal, Bina. "The Gender and Environment Debate: Lessons from India." *Feminist Studies,* 18 (Spring 1992).

Agrawal, Arun. "Indigenous and Scientific Knowledge: Some Critical Comments." *Indigenous Knowledge and Development Monitor* 3 (Dec. 1995).

Ajzenberg-Selove, Fay. *A Matter of Choices: Memoirs of a Female Physicist.* New Brunswick: Rutgers University Press, 1994.

Aldhous, Peter. "Germany: The Backbreaking Work of Scientist-Homemakers." *Science* 263 (11 March 1994).

Alic, Margaret. *Hypatia's Heritage: A History of Women in Science from Antiquity to the Late Nineteenth Century.* London: Women's Press, 1986.

American Association of University Women. *How Schools Shortchange Girls.* Washington: AAUW Educational Foundation, 1992.

Bacus, Elisabeth, et al., eds. *A Gendered Past: A Critical Biography of Gender in Archaeology.* Ann Arbor: University of Michigan Museum of Anthropology, Technical Report 25, 1993.

Badinter, Elisabeth. *Mother Love: Myth and Reality.* New York: Macmillan, 1981.

Balme, Jane, and Wendy Beck, "Archaeology and Feminism: Views on the Origins of the Division of Labour." In *Women in Archaeology,* ed. du Cros and Smith.

Barad, Karen. "A Feminist Approach to Teaching Quantum Physics." In *Teaching the Majority,* ed. Sue Rosser. New York: Teachers College Press, 1995.

Barinaga, Marcia. "Is There a 'Female Style' in Science?" *Science* 260 (16 April 1993).
———. "Profile of a Field: Neuroscience." *Science* 255 (13 March 1992).
———. "Surprises across the Cultural Divide." *Science* 263 (11 March 1994).
Biology and Gender Study Group. "The Importance of Feminist Critique for Contemporary Cell Biology." In *Feminism and Science,* ed. Tuana.
Birke, Lynda. *Women, Feminism, and Biology: The Feminist Challenge.* New York: Methuen, 1986.
Birke, Lynda, and Ruth Hubbard, eds. *Reinventing Biology: Respect for Life and the Creation of Knowledge.* Bloomington: Indiana University Press, 1995.
Blackwell, Elizabeth. *Opening the Medical Profession to Women: Autobiographical Sketches.* 1914; New York: Schocken, 1977.
Bleier, Ruth. "A Decade of Feminist Critiques in the Natural Sciences." *Signs* 14 (1988).
———. *Science and Gender: A Critique of Biology and Its Theories on Women.* New York: Pergamon, 1984.
———, ed. *Feminist Approaches to Science.* New York: Pergamon, 1986.
Brandon, Paul, Barbara Newton, and Ormond Hammond. "Children's Mathematics Achievement in Hawaii: Sex Differences Favoring Girls." *American Educational Research Journal* 24 (1987).
Briscoe, Anne. "Scientific Sexism: The World of Chemistry." In *Women in the Scientific and Engineering Professions,* ed. Violet Haas and Carolyn Perrucci. Ann Arbor: University of Michigan Press, 1984.
Bug, Amy. "Gender and Physical Science: A Hard Look at a Hard Science." In *Women Succeeding in the Sciences: Theories and Practices across the Disciplines,* ed. J. Bart. West Lafayette, Ind.: Purdue University Press, forthcoming.
Clarke, Edward. *Sex in Education: A Fair Chance for Girls.* Boston: Osgood, 1874.
Climbing the Academic Ladder: Doctoral Women Scientists in Academe. Washington: National Academy of Sciences, 1979.
Cole, Jonathan. *Fair Science: Women in the Scientific Community.* New York: Free Press, 1979.
Cole, Jonathan, and Harriet Zuckerman. "Marriage, Motherhood, and Research Performance in Science." 1987; rpt. in *The Outer Circle,* ed. Zuckerman, Cole, and Bruer.
———. "The Productivity Puzzle: Persistence and Change in Patterns of Publication of Men and Women Scientists." In *Advances in Motivation and Achievement,* ed. Steinkamp and Maehr, vol. 2.
Cole, Stephen, and Robert Fiorentine. "Discrimination against Women in Science: The Confusion of Outcome with Process." In *The Outer Circle,* ed. Zuckerman, Cole, and Bruer.
Collins, Patricia Hill. *Black Feminist Thought: Knowledge, Consciousness, and the Politics of Empowerment.* New York: Routledge, 1991.
Committee on Women in Science and Engineering, National Research Council.

Women Scientists and Engineers Employed in Industry: Why So Few? Washington: National Academy Press, 1994.

Conkey, Margaret. "Making the Connections: Feminist Theory and Archaeologies of Gender." In *Women in Archaeology*, ed. du Cros and Smith.

Conkey, Margaret, and Janet Spector. "Archaeology and the Study of Gender." *Advances in Archaeological Method and Theory* 7 (1984).

Conkey, Margaret, with Sarah Williams. "Original Narratives: The Political Economy of Gender in Archaeology." In *Gender at the Crossroads of Knowledge*, ed. di Leonardo.

Cooper, David, Renée Vellvé, and Henk Hobbelink, eds. *Growing Diversity: Genetic Resources and Local Food Security*. London: Intermediate Technology Publication, 1992.

Darwin, Charles. *The Descent of Man, and Selection in Relation to Sex* (1871), ed. John Bonner and Robert May. Princeton: Princeton University Press, 1981.

Davis, Cinda-Sue, Angela Ginorio, Carol Hollenshead, Barbara Lazarus, and Paula Rayman, eds. *The Equity Equation: Fostering the Advancement of Women in the Sciences, Mathematics, and Engineering*. San Francisco: Jossey-Bass, 1996.

Davis, Cinda-Sue, and Sue Rosser. "Program and Curricular Interventions." In *The Equity Equation*, ed. Davis, et al.

di Leonardo, Micaela, ed. *Gender at the Crossroads of Knowledge: Feminist Anthropology in the Postmodern Era*. Berkeley: University of California Press, 1991.

Dix, Linda, ed. *Women: Their Underrepresentation and Career Differentials in Science and Engineering*. Washington: National Academy Press, 1987.

Donnison, Jean. *Midwives and Medical Men: A History of Inter-Professional Rivalries and Women's Rights*. London: Heinemann, 1977.

Doyal, Lesley. *What Makes Women Sick: Gender and the Political Economy of Health*. New Brunswick: Rutgers University Press, 1995.

du Cros, Hilary, and Laurajane Smith, eds. *Women in Archaeology: A Feminist Critique*. Canberra: Department of Prehistory, Australian National University, 1993.

Easlea, Brian. "The Masculine Image of Science with Special Reference to Physics." In *Perspectives on Gender and Science*, ed. Jan Harding. London: Falmer, 1986.

Epstein, Vivian, and Sheldon Epstein. *History of Women in Science for Young People*. Denver: VSE Publisher, 1994.

Etzkowitz, Henry, Carol Kemelgor, Michael Neuschatz, Brian Uzzi, and Joseph Alonzo. "The Paradox of Critical Mass for Women in Science." *Science* 266 (7 Oct. 1994).

Faruqui, Aktar, Mohamed Hassan, and Gabriella Sandri, eds. *The Role of Women in the Development of Science and Technology in the Third World*. Singapore: World Scientific Publishing, 1991.

Fausto-Sterling, Anne. "Life in the XY Corral." *Women's Studies International Forum* 12 (1989).

———. *Myths of Gender: Biological Theories about Women and Men*, 2nd ed. New York: Basic Books, 1992.

Fedigan, Linda. "The Changing Role of Women in Models of Human Evolution." *Annual Review of Anthropology* 15 (1986).

———. "Is Primatology a Feminist Science?" In *Women in Human Evolution*, ed. Hager.

———. "Science and the Successful Female: Why There Are So Many Women Primatologists." *American Anthropologist* 96 (1994).

Fedigan, Linda, and Laurence Fedigan. "Gender and the Study of Primates." In *Gender and Anthropology*, ed. Morgen.

Fennema, Elizabeth, and Gilah Leder, eds. *Mathematics and Gender*. New York: Teachers College Press, 1990.

Findlen, Paula. "Translating the New Science: Women and the Circulation of Knowledge in Enlightenment Italy." *Configurations* 2 (1995).

Forman, Paul. "Behind Quantum Electronics: National Security as Basis for Physical Research in the United States, 1940–1960." *Historical Studies in the Physical and Biological Sciences* 18 (1987).

Galison, Peter. "Fortran, Physics, and Human Nature." In *The Invention of Physical Science*, ed. Mary Jo Nye, Joan Richards, and Roger Steuwer. Dordrecht: Kluwer, 1992.

———. "The Many Faces of Big Science." In *Big Science*, ed. Galison and Hevly.

Galison, Peter, and Bruce Hevly, eds. *Big Science: The Growth of Large-Scale Research*. Stanford: Stanford University Press, 1992.

Gamble, Vanessa, and Bonnie Blustein. "Racial Differentials in Medical Care," In *Women and Health Research*, ed. Mastroianni, Faden, and Federman, vol. 2.

Gero, Joan. "The Social World of Prehistoric Facts: Gender and Power in Paleoindian Research." In *Women in Archaeology*, ed. du Cros and Smith.

Gero, Joan, and Margaret Conkey, eds. *Engendering Archaeology: Women and Prehistory*. Oxford: Blackwell, 1991.

Gilbert, Scott. *Developmental Biology*, 5th ed. Sunderland, Mass.: Sinauer, 1997.

Gowaty, Patricia, ed. *Feminism and Evolutionary Biology: Boundaries, Intersections, and Frontiers*. New York: Chapman and Hall, 1997.

Gross, Paul, and Norman Levitt. *Higher Superstition: The Academic Left and Its Quarrels with Science*. Baltimore: Johns Hopkins University Press, 1994.

Gusterson, Hugh. "Becoming a Weapons Scientist." In *Technoscientific Imaginaries : Conversations, Profiles, and Memoirs*, ed. George Marcus. Chicago, University of Chicago Press, 1995.

Hager, Lori, ed. *Women in Human Evolution*. New York: Routledge, 1997.

Halpern, Diane. *Sex Differences in Cognitive Abilities*. Hillsdale, N.J.: Erlbaum, 1992.

Haramundanis, Katherine, ed. *Cecilia Payne-Gaposchkin: An Autobiography and Other Recollections*. Cambridge: Cambridge University Press, 1984.

Haraway, Donna. *Primate Visions: Gender, Race and Nature in the World of Modern Science*. New York: Routledge, 1989.

———. *Simians, Cyborgs, and Women: The Reinvention of Nature*. New York: Routledge, 1991.

Harding, Sandra. *The Science Question in Feminism.* Ithaca: Cornell University Press, 1986.
———. *Whose Science? Whose Knowledge? Thinking from Women's Lives.* Ithaca: Cornell University Press, 1991.
Harrison, Michelle. "Women's Health: New Models of Care and a New Academic Discipline." *Journal of Women's Health* 2 (1993).
Haseltine, Florence, and Beverly Jacobson, eds. *Women's Health Research: A Medical and Policy Primer.* Washington: Health Press, 1997.
Healy, Bernadine. "Women in Science: From Panes to Ceilings." *Science* 255 (13 March 1992).
———. "Women's Health, Public Welfare." *JAMA* 266 (1991).
Henley, Nancy. *Body Politics: Power, Sex, and Nonverbal Communication.* Englewood Cliffs, N.J.: Prentice Hall, 1977.
Henrion, Claudia. *Women in Mathematics: The Addition of Difference.* Bloomington: Indiana University Press, 1997.
Herzenberg, Caroline, and Ruth Howes. "Women of the Manhattan Project." *Technology Review* 96 (Nov./Dec. 1993).
Hewitt, Nancy, and Elaine Seymour. "Factors Contributing to High Attrition Rates among Science and Engineering Undergraduate Majors." Report to the Alfred P. Sloan Foundation, 26 April 1991.
Hochschild, Arlie. *The Second Shift.* New York: Avon, 1989.
Hrdy, Sarah. "Empathy, Polyandry, and the Myth of the Coy Female." In *Feminist Approaches to Science,* ed. Bleier.
———. *The Woman That Never Evolved.* Cambridge, Mass.: Harvard University Press, 1981.
Hubbard, Ruth. *The Politics of Women's Biology.* New Brunswick: Rutgers University Press, 1990.
———. *Profitable Promises: Essays on Women, Science and Health.* Monroe, Me.: Common Courage Press, 1994.
Hubbard, Ruth, Mary Sue Henifin, and Barbara Fried, eds. *Biological Woman: The Convenient Myth.* Cambridge, Mass.: Schenkman, 1982.
Hull, Gloria, Patricia Bell Scott, and Barbara Smith, eds. *All the Women Are White, All the Blacks Are Men, but Some of Us Are Brave: Black Women's Studies.* Old Westbury, N.Y.: Feminist Press, 1981.
Hyde, Janet, Elizabeth Fennema, and Susan Lamon, "Gender Differences in Mathematics Performance: A Meta-Analysis." *Psychological Bulletin* 107 (1990).
Hyde, Janet, Elizabeth Fennema, Marilyn Ryan, Laurie Frost, and Carolyn Hopp. "Gender Comparisons of Mathematics Attitudes and Affects." *Psychology of Women Quarterly* 15 (1990).
Hynes, Patricia. "Toward a Laboratory of One's Own: Lesbians in Science." In *Lesbian Studies: Present and Future,* ed. Margaret Cruikshank. Old Westbury, N.Y.: Feminist Press, 1982.
"Interview with Andrea Dupree." In *The Outer Circle,* ed. Zuckerman, Cole, and Bruer.

Jasanoff, Sheila, Gerald Markle, James Petersen, and Trevor Pinch, eds. *Handbook of Science and Technology Studies.* Thousand Oaks, Calif.: Sage, 1995.

Johanson, Donald, and Maitland Edey. *Lucy: The Beginnings of Humankind.* New York: Warner, 1981.

Johnson, Tracy, and Elizabeth Fee. "Women's Health Research: An Introduction." In *Women's Health Research,* ed. Haseltine and Jacobson.

Kahle, Jane. "Images of Science: The Physicist and the Cowboy." In *Gender Issues in Science Education,* ed. Barry Fraser and Geoff Giddings. Perth: Curtin University of Technology, 1987.

Kass-Simon, Gabriele, and Patricia Farnes, eds. *Women of Science: Righting the Record.* Bloomington: Indiana University Press, 1990.

Keller, Evelyn Fox. *Refiguring Life: Metaphors of Twentieth-Century Biology.* New York: Columbia University Press, 1995.

———. *Reflections on Gender and Science.* New Haven: Yale University Press, 1985.

———. *Secrets of Life, Secrets of Death: Essays on Language, Gender and Science.* New York: Routledge, 1992.

Keller, Evelyn Fox, and Helen Longino, eds. *Feminism and Science.* Oxford and New York: Oxford University Press, 1996.

Kenschaft, Patricia, and Sandra Keith, eds. *Winning Women into Mathematics.* Washington: Committee on Participation of Women, Mathematical Association of America, 1991.

Kistiakowsky, Vera. "Women in Physics: Unnecessary, Injurious and Out of Place?" *Physics Today* 33 (Feb. 1980).

Kotte, Dieter. *Gender Differences in Science Achievement in Ten Countries.* Frankfurt: Peter Lang, 1992.

Krieger, Nancy, and Elizabeth Fee. "Man-Made Medicine and Women's Health: The Biopolitics of Sex/Gender and Race/Ethnicity," In *Women's Health, Politics, and Power: Essays on Sex/Gender, Medicine, and Public Health,* ed. Elizabeth Fee and Nancy Krieger. Amityville, N.Y.: Baywood, 1994.

Krieger, Nancy, and Sally Zierler. "Accounting for Health of Women." *Current Issues in Public Health* 1 (1995).

Kumagai, Jean. "Survey and Site Visits Evaluate 'Climate' for Women in Physics." *Physics Today* 47 (Aug. 1994).

LaFollette, Marcel. *Making Science Our Own: Public Images of Science, 1910-1955.* Chicago: University of Chicago Press, 1990.

Lambert, Anne-Thérèse de Marguenat de Courcelles, marquise de. *Reflections nouvelles sur les femmes.* 1727; London, 1820.

Laqueur, Thomas. *Making Sex: Body and Gender from the Greeks to Freud.* Cambridge, Mass.: Harvard University Press, 1990.

Lee, Richard, and Irven DeVore, eds. *Man the Hunter.* Chicago: Aldine, 1968.

Lerner, Gerda. *The Creation of Feminist Consciousness: From the Middle Ages to Eighteen-Seventy.* New York: Oxford University Press, 1993.

Long, J. Scott. "Measures of Sex Differences in Scientific Productivity." *Social Forces* 71 (1992).

Longino, Helen. "Cognitive and Non-Cognitive Values in Science: Rethinking the Dichotomy." In *Feminism, Science, and the Philosophy of Science*, ed. Nelson and Nelson.
——. *Science as Social Knowledge: Values and Objectivity in Scientific Inquiry.* Princeton: Princeton University Press, 1990.
——. "Subjects, Power, and Knowledge: Description and Prescription in Feminist Philosophies of Science." In *Feminism and Science*, ed. Keller and Longino.
Lougee, Carolyn. *Le Paradis des Femmes: Women, Salons, and Social Stratification in Seventeenth-Century France.* Princeton: Princeton University Press, 1976.
Lowe, Marian, and Ruth Hubbard, eds. *Woman's Nature: Rationalizations of Inequality.* Elmsford, N.Y.: Pergamon, 1983.
Margenau, Henry, David Bergamini, and the Editors of LIFE, eds. *The Scientist.* New York: Time, Inc., 1964.
Marland, Hilary, ed. *The Art of Midwifery: Early Modern Midwives in Europe.* London: Routledge, 1993.
Martin, Emily. "The Egg and the Sperm: How Science has constructed a Romance Based on Stereotypical Male-Female Roles." *Signs* 16 (1991).
——. *The Woman in the Body: A Cultural Analysis of Reproduction,* 2nd ed. Boston: Beacon, 1992.
Mastroianni, Anna, Ruth Faden, and Daniel Federman, eds. *Women and Health Research,* vol. 2. Washington: National Academy Press, 1994.
Matyas, Marsha, and Shirley Malcom, eds. *Investing in Human Potential: Science and Engineering at the Crossroads.* Washington: American Association for the Advancement of Science, 1991.
McGrayne, Sharon. *Nobel Prize Women in Science: Their Lives, Struggles, and Momentous Discoveries.* Secaucus, N.J.: Carol, 1993.
Mendelsohn, Kathleen, Linda Nieman, Krista Isaacs, Sophia Lee, and Sandra Levison. "Sex and Gender Bias in Anatomy and Physical Diagnosis Text Illustrations." *JAMA* 262 (26 Oct. 1994).
Merchant, Carolyn. *Earthcare: Women and the Environment.* New York: Routledge, 1995.
Messer-Davidow, Ellen, David Shumway, and David Sylvan, eds. *Knowledges: Historical and Critical Studies in Disciplinarity.* Charlottesville: University Press of Virginia, 1993.
Morbeck, Mary, Alison Galloway, and Adrienne Zihlman, eds. *The Evolving Female: A Life-History Perspective.* Princeton: Princeton University Press, 1997.
Morgen, Sandra, ed. *Gender and Anthropology: Critical Reviews for Research and Teaching.* Washington: American Anthropological Association, 1989.
Morrow, Charlene, and James Morrow, "Whose Math Is It, Anyway?" *Initiatives* 55, 3 (1993).
Narrigan, Deborah, Jane Zones, Nancy Worchester, and Maxine Jo Grad, "Research to Improve Women's Health: An Agenda for Equity." In *Women's Health,* ed. Ruzek, Olesen, and Clarke.

National Science Foundation. *Characteristics of Doctoral Scientists and Engineers in the United States: 1995.* Arlington, Va., 1997.

———. *Women and Minorities in Science and Engineering.* Washington, 1990.

———. *Women and Minorities in Science and Engineering: An Update.* Washington, 1992.

———. *Women, Minorities, and Persons with Disabilities in Science and Engineering: 1994.* Arlington, Va., Nov. 1994.

———. *Women, Minorities, and Persons with Disabilities in Science and Engineering: 1996.* Arlington, Va., Sept. 1996.

Nechas, Eileen, and Denise Foley. *Unequal Treatment: What You Don't Know about How Women Are Mistreated by the Medical Community.* New York: Simon and Schuster, 1994.

Nelson, Lynn, and Jack Nelson. *Feminism, Science, and the Philosophy of Science.* Dordrecht: Kluwer, 1996.

Nelson, Margaret, Sarah Nelson, and Alison Wylie, eds. *Equity Issues for Women in Archeology.* Arlington, Va.: American Anthropological Association: Archeological Papers, no. 5, 1994.

Niedersächsisches Ministerium für Wissenschaft und Kultur. *Berichte aus der Frauenforschung: Perspektiven für Naturwissenschaften, Technik und Medizin.* Hannover, 1997.

Oelsner, Elise. *Die Leistungen der deutschen Frau in der letzten vierhundert Jahren auf wissenschaftlichen Gebiete.* Guhrau, 1894.

Office of Research on Women's Health. *Report of the National Institutes of Health: Opportunities for Research on Women's Health.* Bethesda: National Institutes of Health, 1991.

Oudshoorn, Nelly. *Beyond the Natural Body: An Archeology of Sex Hormones.* London: Routledge, 1994.

Pearson, Willie, Jr. *Black Scientists, White Society, and Colorless Science.* New York: Associated Faculty Press, 1985.

Pizan, Christine de. *The Book of the City of Ladies* (1405), trans. Earl Jeffrey Richards. New York: Persea, 1982.

Primmer, Lesley. "Women's Health Research: Congressional Action and Legislative Gains: 1990-1994." In *Women's Health Research,* ed. Haseltine and Jacobson.

Proctor, Robert. *Value-Free Science? Purity and Power in Modern Knowledge.* Cambridge, Mass.: Harvard University Press, 1991.

Rayman, Paula, and Belle Brett. *Pathways for Women in the Sciences.* Wellesley, Mass.: Wellesley College Center for Research on Women, 1993.

Rebière, Alphonse. *Les Femmes dans la science.* 2d ed. Paris, 1897.

Rose, Hilary. "Hand, Brain, and Heart: A Feminist Epistemology for the Natural Sciences." *Signs* 9 (1983).

———. *Love, Power, and Knowledge: Towards a Feminist Transformation of the Sciences.* Bloomington: Indiana University Press, 1994.

Ross, Andrew, ed. *Science Wars.* Durham: Duke University Press, 1996.

Rosser, Phyllis. *The SAT Gender Gap: ETS Responds, A Research Update.* Washington: Center for Women Policy Studies, 1992.

———. *The SAT Gender Gap: Identifying the Causes.* Washington: Center for Women Policy Studies, 1989.

Rosser, Sue. *Female-Friendly Science: Applying Women's Studies Methods and Theories to Attract Students.* New York: Pergamon, 1990.

———. *Re-Engineering Female Friendly Science.* New York: Teachers College Press, 1997.

———. *Women's Health: Missing from U.S. Medicine.* Bloomington, Indiana University Press, 1994.

———, ed. *Teaching the Majority: Breaking the Gender Barrier in Science, Mathematics, and Engineering.* New York: Teachers College Press, 1995.

Rossiter, Margaret. "The [Matthew] Matilda Effect in Science." *Social Studies of Science* 23 (1993).

———. *Women Scientists in America: Before Affirmative Action, 1940–1972.* Baltimore: Johns Hopkins University Press, 1995.

———. *Women Scientists in America: Struggles and Strategies to 1940.* Baltimore: Johns Hopkins Press, 1982.

Rousseau, Jean-Jacques. *Lettre à M. d'Alembert sur les spectacles* (1758), ed. L. Brunel. Paris, 1896.

Rowell, Thelma. "Introduction: Mothers, Infants, and Adolescents." In *Female Primates: Studies by Women Primatologists,* ed. Meredith Small, 13–16. New York: Alan Liss, 1984.

Russett, Cynthia. *Sexual Science: The Victorian Construction of Womanhood.* Cambridge, Mass.: Harvard University Press, 1989.

Ruzek, Sheryl, Adele Clarke, and Virginia Olesen, "Social, Biomedical, and Feminist Models of Women's Health." In *Women's Health,* ed. Ruzek, Olesen, and Clarke.

Ruzek, Sheryl, Virginia Olesen, and Adele Clarke, eds. *Women's Health: Complexities and Differences.* Columbus: Ohio State University Press, 1997.

Schiebinger, Londa. "Creating Sustainable Science." In *Women, Gender, and Science: New Directions,* ed. Sally Kohlstedt and Helen Longino. *Osiris* 12 (1997).

———. "Lost Knowledge, Bodies of Ignorance, and the Poverty of Taxonomy as Illustrated by the Curious Fate of *Flos Pavonis,* an Abortifacient." In *Picturing Science, Producing Art,* ed. Caroline Jones and Peter Galison. New York: Routledge, 1998.

———. *The Mind Has No Sex? Women in the Origins of Modern Science.* Cambridge, Mass.: Harvard University Press, 1989.

———. *Nature's Body: Gender in the Making of Modern Science.* Boston: Beacon, 1993.

Sebrechts, Jadwiga. "Cultivating Scientists at Women's Colleges." *Initiatives* 55, 2 (1993).

Shiva, Vandana. *Staying Alive: Women, Ecology and Development.* London: Zed, 1988.

Sime, Ruth. *Lise Meitner: A Life in Physics.* Berkeley: University of California Press, 1996.

Snow, C. P. *The Two Cultures and the Scientific Revolution*. New York: Cambridge University Press, 1961.

Sonnert, Gerhard, and Gerald Holton. *Gender Differences in Science Careers: The Project Access Study*. New Brunswick: Rutgers University Press, 1995.

————. " 'Glass Ceiling' vs. 'Threshold': The Career Paths of Women Scientists." Paper presented at the Society for Social Studies of Science Annual Meeting, Cambridge, Mass., 1991.

————. *Who Succeeds in Science? The Gender Dimension*. New Brunswick: Rutgers University Press, 1995.

Spanier, Bonnie. *Im/partial Science: Gender Ideology in Molecular Biology*. Bloomington: Indiana University Press, 1995.

Spector, Barbara. "Women Astronomers Say Discrimination in Field Persists." *Scientist 5* (1 April 1991).

Spertus, Ellen. "Why Are There So Few Female Computer Scientists?" Report #AIM 1315. Cambridge, Mass.: Artificial Intelligence Laboratory, 1991.

Squier, Susan. *Babies in Bottles: Twentieth-Century Visions of Reproductive Technology*. New Brunswick: Rutgers University Press, 1994.

Stamps, Judy. "The Role of Females in Extrapair Copulations." In *Feminism and Evolutionary Biology*, ed. Gowaty.

Steinkamp, Marjorie, and Martin Maehr, eds. *Advances in Motivation and Achievement: Women in Science*. Greenwich, Conn.: JAI Press, 1984.

Storer, Norman. "The Hard Sciences and the Soft." *Bulletin of the Medical Library Association* 55 (1967).

Stricker, Lawrence, Donald Rock, and Nancy Burton. *Sex Difference in SAT Predictions of College Grades*. New York: College Board, 1991.

Strum, Shirley, and Linda Fedigan. "Theory, Method and Gender: What Changed Our Views of Primate Society?" In *The New Physical Anthropology*, ed. Shirley Strum, Donald Lindburg, and David Hamburg. Upper Saddle River, N.J.: Prentice Hall, 1999.

Tannen, Deborah. *You Just Don't Understand: Women and Men in Conversation*. New York: Ballantine, 1990.

Terry, Jennifer, and Jacqueline Urla, eds. *Deviant Bodies: Critical Perspectives on Difference in Science and Popular Culture*. Bloomington: Indiana University Press, 1995.

Traweek, Sharon. *Beamtimes and Lifetimes: The World of High Energy Physicists*. Cambridge, Mass.: Harvard University Press, 1988.

————. "Big Science and Colonialist Discourse: Building High-Energy Physics in Japan," In *Big Science*, ed. Galison and Hevly.

Tuana, Nancy, ed. *Feminism and Science*. Bloomington: Indiana University Press, 1989.

Vetter, Betty. *Professional Women and Minorities*. Washington: Commission on Professionals in Science and Technology, Jan. 1994.

————. "What Is Holding up the Glass Ceiling? Barriers to Women in the Science and Engineering Workforce." Occasional Paper 92–3. Washington: Commission on Professionals in Science and Technology, 1992.

Wallis, Lila. "Why a Curriculum in Women's Health." *Journal of Women's Health* 2 (1993).

Watson, James. *The Double Helix: A Personal Account of the Discovery of the Structure of DNA.* 1968; New York: Norton, 1980.

Widnall, Sheila. "AAAS Presidential Lecture: Voices from the Pipeline." *Science* 241 (30 Sept. 1988).

Wolpert, Lewis, and Alison Richards, eds. *A Passion for Science.* Oxford: Oxford University Press, 1988.

Wylie, Alison, "The Engendering of Archaeology: Refiguring Feminist Science Studies." In *Women, Gender, and Science: New Directions,* ed. Sally Kohlstedt and Helen Longino. *Osiris* 12 (1997).

Zappert, Laraine, and Kendyll Stanbury. "In the Pipeline: A Comparative Analysis of Men and Women in Graduate Programs in Science, Engineering, and Medicine at Stanford University." Working Paper 20, Institute for Research on Women and Gender, Stanford University, 1984.

Zihlman, Adrienne. "Misreading Darwin on Reproduction: Reductionism in Evolutionary Theory." In *Conceiving the New World Order: The Global Politics of Reproduction,* ed. Faye Ginsburg and Rayna Rapp. Berkeley: University of California Press, 1995.

———. "The Paleolithic Glass Ceiling: Women in Human Evolution." In *Women in Human Evolution,* ed. Hager.

———. "Sex, Sexes, and Sexism in Human Origins." *Yearbook of Physical Anthropology* 30 (1987).

Zuckerman, Harriet. "The Careers of Men and Women Scientists: A Review of Current Research." In *The Outer Circle,* ed. Zuckerman, Cole, and Bruer, 27–56.

Zuckerman, Harriet, Jonathan Cole, and John Bruer, eds. *The Outer Circle: Women in the Scientific Community.* New Haven: Yale University Press, 1992.

Index

CPSIA information can be obtained at www.ICGtesting.com
Printed in the USA
BVOW06s0000111215

429995BV00002B/2/P